Open Quantum Physics and Environmental Heat Conversion Into Usable Energy

(Volume 3)

Authored by

Eliade Stefanescu

Center of Advanced Studies in Physics of the Romanian Academy
Academy of Romanian Scientists
Bucharest
Romania

Open Quantum Physics and Environmental Heat Conversion into Usable Energy

Volume # 3

Author: Eliade Stefanescu

ISSN (Online): 2542-5072

ISSN (Print): 2542-5064

ISBN (Online): 978-981-5051-09-4

ISBN (Print): 978-981-5051-10-0

ISBN (Paperback): 978-981-5051-11-7

BENTHAM SCIENCE PUBLISHERS LTD.

End User License Agreement (for non-institutional, personal use)

This is an agreement between you and Bentham Science Publishers Ltd. Please read this License Agreement carefully before using the ebook/echapter/ejournal (**"Work"**). Your use of the Work constitutes your agreement to the terms and conditions set forth in this License Agreement. If you do not agree to these terms and conditions then you should not use the Work.

Bentham Science Publishers agrees to grant you a non-exclusive, non-transferable limited license to use the Work subject to and in accordance with the following terms and conditions. This License Agreement is for non-library, personal use only. For a library / institutional / multi user license in respect of the Work, please contact: permission@benthamscience.net.

Usage Rules:

1. All rights reserved: The Work is the subject of copyright and Bentham Science Publishers either owns the Work (and the copyright in it) or is licensed to distribute the Work. You shall not copy, reproduce, modify, remove, delete, augment, add to, publish, transmit, sell, resell, create derivative works from, or in any way exploit the Work or make the Work available for others to do any of the same, in any form or by any means, in whole or in part, in each case without the prior written permission of Bentham Science Publishers, unless stated otherwise in this License Agreement.
2. You may download a copy of the Work on one occasion to one personal computer (including tablet, laptop, desktop, or other such devices). You may make one back-up copy of the Work to avoid losing it.
3. The unauthorised use or distribution of copyrighted or other proprietary content is illegal and could subject you to liability for substantial money damages. You will be liable for any damage resulting from your misuse of the Work or any violation of this License Agreement, including any infringement by you of copyrights or proprietary rights.

Disclaimer:

Bentham Science Publishers does not guarantee that the information in the Work is error-free, or warrant that it will meet your requirements or that access to the Work will be uninterrupted or error-free. The Work is provided "as is" without warranty of any kind, either express or implied or statutory, including, without limitation, implied warranties of merchantability and fitness for a particular purpose. The entire risk as to the results and performance of the Work is assumed by you. No responsibility is assumed by Bentham Science Publishers, its staff, editors and/or authors for any injury and/or damage to persons or property as a matter of products liability, negligence or otherwise, or from any use or operation of any methods, products instruction, advertisements or ideas contained in the Work.

Limitation of Liability:

In no event will Bentham Science Publishers, its staff, editors and/or authors, be liable for any damages, including, without limitation, special, incidental and/or consequential damages and/or damages for lost data and/or profits arising out of (whether directly or indirectly) the use or inability to use the Work. The entire liability of Bentham Science Publishers shall be limited to the amount actually paid by you for the Work.

General:

1. Any dispute or claim arising out of or in connection with this License Agreement or the Work (including non-contractual disputes or claims) will be governed by and construed in accordance with the laws of Singapore. Each party agrees that the courts of the state of Singapore shall have exclusive jurisdiction to settle any dispute or claim arising out of or in connection with this License Agreement or the Work (including non-contractual disputes or claims).
2. Your rights under this License Agreement will automatically terminate without notice and without the

Bentham Science Publishers Pte. Ltd.
80 Robinson Road #02-00
Singapore 068898
Singapore
Email: subscriptions@benthamscience.net

BENTHAM
SCIENCE

CONTENTS

PREFACE

In the first two volumes, we approached the openness of a physical system, from coupling to a complex dissipative environment of Fermions, Bosons, and a free electromagnetic field. Essentially, the open description of a system of interest starts with the dynamics of the total system, including the environment, and consists of the reduction of the system dynamics on the environmental coordinates to dissipative dynamics, with coefficients depending on the coupling matrix elements, the densities of the environmental states, and the occupation probabilities of these states, as a function of temperature. Various empirical descriptions of coupling with the environment violated fundamental principles of quantum mechanics as the uncertainty principle, the zero-point vibration, and the positivity of the density matrix. A positive application of the dissipative dynamics was discovered in the seventies by Lindblad, but Lindblad's master equation became a popular tool, especially in nuclear physics, only in the eighties, when Sandulescu and Scutaru applied this equation to deep inelastic collisions of heavy ions. However, this equation, much used today, is very unsatisfactory, being a phenomenological one, with terms for all the system operators, with unspecified dissipation coefficients. Consequently, in the nineties, and the early 2000s, I obtained master equations for Fermions, Bosons, and a coherent electromagnetic field, with explicit microscopic coefficients for the dissipative coupling with other Fermions, Bosons, and the free electromagnetic field, and terms for non-Markovian effects. Based on these equations, I showed that the entropy of a matter-field system could spontaneously decrease, not only increase as it is asserted by the second law of thermodynamics, for molecular systems. In this framework, I invented a semiconductor device converting environmental heat into usable energy. A theoretical description of this device, and of the quantum mechanical and statistical fundamentals are the objects of the first two volumes. However, in the approach of these fundamentals, I found that a general solution of the Schrödinger equation in the coordinate space does not correctly describe the particle dynamics, according to the Hamilton equations. A correct description is obtained only with propagation wave functions when the Hamiltonian of the time-dependent phase is replaced by the Lagrangian. In this volume, with the relativistic Lagrangian, for a quantum particle, I obtain a more physical description, as an invariant quantity of matter propagating in space, with the mass determined by the dynamic characteristics of the matter density. Quantum mechanics is obtained from the general theory of relativity. I use the formalism of Dirac, who was the big architect of quantum mechanics, and the general theory of relativity. In the whole universe, where it is curved on other dimensions, I regard our four-dimensional physical universe to be an open system, describing the inertial-gravitational dynamics.

CONSENT FOR PUBLICATION

Not applicable.

CONFLICT OF INTEREST

The author declares that he has no affiliation with any organization or entity from a financial point of view in the subject matter or materials discussed in this book.

ACKNOWLEDGEMENT

Declared none.

Eliade Stefanescu

Center of Advanced Studies in Physics of the Romanian Academy
Academy of Romanian Scientists
Bucharest
Romania

DEDICATION

I dedicate this book
to my grandson **Alex** and my granddaughter **Julia**
and to all young people tending to really understanding the world they live in

CHAPTER 1

Introduction

Abstract: We consider the two basic theories of the standard physics: (1) the theory of relativity, as a four-dimensional time-space description of our universe, with the invariance of the time-space interval which leads to the Lorentz transformation, and (2) quantum mechanics, based on the Schrödinger equation, with a solution called wave function. However, we find that this equation is very unsatisfactory, describing fully unlocalized free particles, any initially localized particle rapidly spreading in space, in disagreement with the fundamental Hamilton equations. In agreement with the Hamilton equations, we describe a quantum particle by propagation wave functions in the coordinate and momentum spaces, with the Lagrangian in the time dependent phase instead of the Hamiltonian coming from the Schrödinger equation. With the relativistic Lagrangian, we obtain wave functions describing invariant distributions propagating in space.

Keywords: Light velocity, Time-space interval, Lorentz transformation, Hamiltonian, Kinetic energy, Potential energy, Mass, Schrödinger equation, Operator, Heisenberg picture, Schrödinger picture, Hamilton equation, Wave function, Wave packet, Group velocity, Standard deviation, Mean value, Commutation, Lagrangian, Proper time, Momentum, Conjugate space, Metric tensor.

The standard physics is based on two different descriptions:

(1) the theory of relativity [1], which we consider here in Dirac's formalism [2], as an analytic description of a continuous distribution of matter in a time-space system of coordinates

$$x=\left(x^{\alpha}\right)=\left(x^{0}=ct,x^{1},x^{2},x^{3}\right)=\left(x^{0}=ct,x^{i}\right), \quad \textbf{(1.1)}$$

where c is a universal constant, called the light velocity, and

(2) quantum mechanics [3], as a statistical description of a system punctual particles, any particle randomly having different coordinates $\vec{r}=x\vec{1}_x+y\vec{1}_y+z\vec{1}_z$, with probabilities determined by a wave function $\psi\left(t,\vec{r}\right)$.

The theory of relativity is based on the invariance of the time-space interval, which in two orthogonal systems S and S' of a flat space is

$$ds^2=dx^{0^2}-dx^{1^2}-dx^{2^2}-dx^{3^2}=ds'^2=dx^{0'^2}-dx^{1'^2}-dx^{2'^2}-dx^{3'^2}, \quad \textbf{(1.2)}$$

Eliade Stefanescu

where the power indices of the variables are attached to the contravariant indices of these variables. For the simpler case of two systems S and S' with parallel axes, moving one another in the direction $x^1 \square x^{1'}$, the coordinates x^α of the system S are functions of the coordinates $x^{\alpha'}$ of the system S',

$$\begin{cases} x^0 = x^0\left(x^{0'}, x^{1'}\right) \\ x^1 = x^1\left(x^{0'}, x^{1'}\right) \\ x^2 = x^{2'} \\ x^3 = x^{3'}. \end{cases} \tag{1.3}$$

In this case, from the invariance relation (1.2) with (1.3),

$$\begin{aligned} ds^2 &= dx^{0^2} - dx^{1^2} - dx^{2^2} - dx^{3^2} \\ &= \left({x^0}_{,0'}dx^{0'} + {x^0}_{,1'}dx^{1'}\right)^2 - \left({x^1}_{,0'}dx^{0'} + {x^1}_{,1'}dx^{1'}\right)^2 - dx^{2'^2} - dx^{3'^2} \\ &= dx^{0'^2} - dx^{1'^2} - dx^{2'^2} - dx^{3'^2}, \end{aligned}$$

we obtain the system of equations:

$$\begin{cases} {x^0}_{,0'}^{\,2} - {x^1}_{,0'}^{\,2} = 1 \\ {x^0}_{,1'}^{\,2} - {x^1}_{,1'}^{\,2} = -1 \\ {x^0}_{,0'}{x^0}_{,1'} - {x^1}_{,0'}{x^1}_{,1'} = 0. \end{cases} \tag{1.4}$$

For the differentials of the transformation equations (1.3),

$$\begin{cases} dx^0 = {x^0}_{,0'}dx^{0'} + {x^0}_{,1'}dx^{1'} \\ dx^1 = {x^1}_{,0'}dx^{0'} + {x^1}_{,1'}dx^{1'}, \end{cases}$$

with the first two equations (1.4) we obtain the equations

$$\begin{cases} dx^0 = \sqrt{1 + {x^1}_{,0'}^{\,2}}\,dx^{0'} + {x^0}_{,1'}dx^{1'} \\ dx^1 = {x^1}_{,0'}dx^{0'} + \sqrt{1 + {x^0}_{,1'}^{\,2}}\,dx^{1'}. \end{cases} \tag{1.5}$$

At the same time, from the third equation (1.4) with the other two ones, we obtain the relation

$$x^{0}_{,1'} = x^{1}_{,0'}\frac{x^{1}_{,1'}}{x^{0}_{,0'}} = x^{1}_{,0'}\frac{\sqrt{1+x^{0}_{,1'}{}^{2}}}{\sqrt{1+x^{1}_{,0'}{}^{2}}},$$

where the two functions can be separated:

$$\sqrt{\frac{1}{x^{0}_{,1'}{}^{2}}+1} = \sqrt{\frac{1}{x^{1}_{,0'}{}^{2}}+1}.$$

We obtain $x^{0}_{,1'} = x^{1}_{,0'}$, as equations (1.5) take a form depending on a single function, $x^{1}_{,0'}$,

$$\begin{cases} dx^{0} = \sqrt{1+x^{1}_{,0'}{}^{2}}\,dx^{0'} + x^{1}_{,0'}dx^{1'} = \sqrt{1+x^{1}_{,0'}{}^{2}}\left(dx^{0'} + \dfrac{x^{1}_{,0'}}{\sqrt{1+x^{1}_{,0'}{}^{2}}}dx^{1'}\right) \\ dx^{1} = x^{1}_{,0'}dx^{0'} + \sqrt{1+x^{1}_{,0'}{}^{2}}\,dx^{1'} = \sqrt{1+x^{1}_{,0'}{}^{2}}\left(dx^{1'} + \dfrac{x^{1}_{,0'}}{\sqrt{1+x^{1}_{,0'}{}^{2}}}dx^{0'}\right). \end{cases} \quad \textbf{(1.6)}$$

With the velocity

$$v = \frac{x^{1}_{,0'}}{\sqrt{1+x^{1}_{,0'}{}^{2}}},$$

the coefficient of equations (1.6) takes a form

$$\sqrt{1+x^{1}_{,0'}{}^{2}} = \frac{1}{\sqrt{1-\dfrac{x^{1}_{,0'}{}^{2}}{1+x^{1}_{,0'}{}^{2}}}} = \frac{1}{\sqrt{1-v^{2}}},$$

which depends on this velocity, as equations take the form of the Lorentz transformation:

$$\begin{cases} \mathrm{d}x^0 = \dfrac{\mathrm{d}x^{0'} + v\mathrm{d}x^{1'}}{\sqrt{1-v^2}} \\ \mathrm{d}x^1 = \dfrac{\mathrm{d}x^{1'} + v\mathrm{d}x^{0'}}{\sqrt{1-v^2}}. \end{cases} \quad \textbf{(1.7)}$$

From these homogeneous equations we notice that the time coordinate x^0 and the spatial coordinates x^i are similar time-space coordinates which satisfy the invariance relation (1.2).

At the same time, in the standard quantum mechanics, a quantum particle with a mass M and a momentum $\vec{p}$, in an energy potential $U(\vec{r})$, is described by a wave function $\psi(t,\vec{r})$. In agreement with the classical mechanics, with the Hamiltonian $H(\vec{r},\vec{p})$ as the sum of the kinetic energy $T(\vec{p})$ with the potential energy $U(\vec{r})$,

$$H(\vec{r},\vec{p}) = T(\vec{p}) + U(\vec{r}) = \frac{\vec{p}^2}{2M} + U(\vec{r}), \quad \textbf{(1.8)}$$

this wave function is considered as a solution of the Schrödinger equation [3],

$$\mathrm{i}\hbar\frac{\partial}{\partial t}\psi(\vec{r},t) = H(\vec{r},\vec{p})\psi(\vec{r},t). \quad \textbf{(1.9)}$$

With the Hamiltonian $H(\vec{r},\vec{p})$ as an operator

$$H(\vec{r},\vec{p}) = \mathrm{i}\hbar\frac{\partial}{\partial t}, \quad \textbf{(1.10)}$$

and the momentum operator

$$\vec{p} = -\mathrm{i}\hbar\frac{\partial}{\partial \vec{r}}, \quad \textbf{(1.11)}$$

the Schrödinger equation (1.9) with the Hamiltonian (1.8) takes the explicit form

$$\mathrm{i}\hbar\frac{\partial}{\partial t}\psi(\vec{r},t) = \left[-\frac{\hbar^2}{2M}\frac{\partial^2}{\partial \vec{r}^2} + U(\vec{r})\right]\psi(\vec{r},t). \quad \textbf{(1.12)}$$

With the energy E as the Hamiltonian eigenvalue, this equation takes a time independent form

$$H(\vec{r},\vec{p})\psi(\vec{r}) = \left[-\frac{\hbar^2}{2M}\frac{\partial^2}{\partial\vec{r}^2} + U(\vec{r})\right]\psi(\vec{r}) = E\psi(\vec{r}), \quad \textbf{(1.13)}$$

used in numerous applications of atomic, nuclear, and condensed matter physics [3 - 8]. The solution of the time-dependent Schrödinger equation (1.9) is of the form

$$|\psi(\vec{r},t)\rangle = e^{-\frac{i}{\hbar}H(\vec{r},\vec{p})t}|\psi_0(\vec{r})\rangle \quad \textbf{(1.14)}$$

described by an evolution operator

$$U_t = e^{-\frac{i}{\hbar}H(\vec{r},\vec{p})t}. \quad \textbf{(1.15)}$$

With the wave function (1.14), the mean value of an operator A takes the form

$$\begin{aligned}\langle\psi(\vec{r},t)|A|\psi(\vec{r},t)\rangle &= \langle\psi_0(\vec{r})|e^{\frac{i}{\hbar}H(\vec{r},\vec{p})t}Ae^{-\frac{i}{\hbar}H(\vec{r},\vec{p})t}|\psi_0(\vec{r})\rangle \\ &= \langle\psi_0(\vec{r})|A(t)|\psi_0(\vec{r})\rangle\end{aligned}$$

depending on the time dependent operator

$$A(t) = e^{\frac{i}{\hbar}H(\vec{r},\vec{p})t}Ae^{-\frac{i}{\hbar}H(\vec{r},\vec{p})t}, \quad \textbf{(1.16)}$$

and the time independent wave function $|\psi_0(\vec{r})\rangle$, in the Heisenberg picture, as the time dependent wave function $|\psi(\vec{r},t)\rangle$ and the time independent operator A are in the Schrödinger picture. From (1.16) we obtain the Heisenberg dynamic equation:

$$\frac{\partial}{\partial t}A(t) = \frac{i}{\hbar}\left[H(\vec{r},\vec{p}),A(t)\right]. \quad \textbf{(1.17)}$$

With the momentum operator (1.11) we obtain the commutation relation

$$\begin{aligned}
[\vec{r},\vec{p}] &= -\mathrm{i}\hbar\vec{r}\frac{\partial}{\partial\vec{r}} + \mathrm{i}\hbar\frac{\partial}{\partial\vec{r}}\vec{r} = -\mathrm{i}\hbar\vec{r}\frac{\partial}{\partial\vec{r}} + \mathrm{i}\hbar\vec{r}\frac{\partial}{\partial\vec{r}} + \mathrm{i}\hbar\frac{\partial\vec{r}}{\partial\vec{r}} \\
&= \mathrm{i}\hbar\left(\vec{1}_x\frac{\partial}{\partial x} + \vec{1}_y\frac{\partial}{\partial y} + \vec{1}_x\frac{\partial}{\partial z}\right)\left(\vec{1}_x x + \vec{1}_y y + \vec{1}_x z\right) \\
&= \mathrm{i}\hbar\left(\frac{\partial x}{\partial x} + \frac{\partial y}{\partial y} + \frac{\partial z}{\partial z}\right) = 3\mathrm{i}\hbar.
\end{aligned} \qquad \textbf{(1.18)}$$

With this relation, from the Heisenberg equation (1.17) we obtain mean value equations

$$\begin{aligned}
\frac{\partial}{\partial t}\langle\vec{r}(t)\rangle &= \frac{\mathrm{i}}{\hbar}\left\langle\left[\frac{\vec{p}^2}{2M} + U(\vec{r}), \vec{r}\right]\right\rangle = \frac{\mathrm{i}}{2M\hbar}\langle\vec{p}[\vec{p},\vec{r}] + [\vec{p},\vec{r}]\vec{p}\rangle \\
&= 3\frac{\langle\vec{p}(t)\rangle}{M} \\
\frac{\partial}{\partial t}\langle\vec{p}(t)\rangle &= \frac{\mathrm{i}}{\hbar}\left\langle\left[\frac{\vec{p}^2}{2M} + U(\vec{r}), \vec{p}\right]\right\rangle = \frac{\mathrm{i}}{\hbar}\left\langle\left[U(\vec{r}), -\mathrm{i}\hbar\frac{\partial}{\partial\vec{r}}\right]\right\rangle \\
&= \frac{\mathrm{i}}{\hbar}\left\langle -\mathrm{i}\hbar U(\vec{r})\frac{\partial}{\partial\vec{r}} + \mathrm{i}\hbar\frac{\partial}{\partial\vec{r}}U(\vec{r})\right\rangle = \frac{\mathrm{i}}{\hbar}\left\langle -\mathrm{i}\hbar U(\vec{r})\frac{\partial}{\partial\vec{r}} + \mathrm{i}\hbar\frac{\partial U(\vec{r})}{\partial\vec{r}} + \mathrm{i}\hbar U(\vec{r})\frac{\partial}{\partial\vec{r}}\right\rangle \\
&= -\left\langle\frac{\partial U(\vec{r})}{\partial\vec{r}}\right\rangle,
\end{aligned} \qquad \textbf{(1.19)}$$

which can be compared with the fundamental Hamilton equations

$$\begin{aligned}
\frac{\mathrm{d}\vec{r}}{\mathrm{d}t} &= \frac{\partial}{\partial\vec{p}}H(\vec{r},\vec{p}) = \frac{\vec{p}}{M} \\
\frac{\mathrm{d}\vec{p}}{\mathrm{d}t} &= -\frac{\partial}{\partial\vec{r}}H(\vec{r},\vec{p}) = -\frac{\partial U(\vec{r})}{\partial\vec{r}}.
\end{aligned} \qquad \textbf{(1.20)}$$

We notice that although the second equation (1.19) is in agreement with the second Hamilton equation (1.20), the first mean-value equation (1.19) differs from the first Hamilton equation (1.20) with a factor 3. More than that, for a free particle, $U(\vec{r}) = 0$, the solution of the Schrödinger equation (1.12) takes the form of a plane wave, which means that the mean value $\langle\vec{r}(t)\rangle$ of the first equation (1.19) does not make any sense. This means that for the Schrödinger equation (1.9) we need to consider a solution of the form of propagation wave function,

$$\psi(\vec{r},t) = \frac{1}{(2\pi\hbar)^{3/2}} \int \varphi(\vec{p},t) e^{\frac{i}{\hbar}[\vec{p}\vec{r} - H(\vec{r},\vec{p})t]} \mathrm{d}^3\vec{p}, \tag{1.21}$$

which in fact stands at the basis of the derivation of the Schrödinger equation [3].With the inverse Fourier transform

$$\varphi(\vec{p},t) = \frac{1}{(2\pi\hbar)^{3/2}} \int \psi(\vec{r},t) e^{-\frac{i}{\hbar}[\vec{p}\vec{r} - H(\vec{r},\vec{p})t]} \mathrm{d}^3\vec{r}, \tag{1.22}$$

we obtain a description of the quantum particle in the two conjugate spaces, of the coordinates and of the momentum. In this case, the dynamics of the system is described by the group velocities of the two wave packets:

$$\begin{aligned} \frac{\mathrm{d}\vec{r}}{\mathrm{d}t} &= \frac{\partial}{\partial\vec{p}} H(\vec{r},\vec{p}) = \frac{\partial T(\vec{p})}{\partial\vec{p}} = \frac{\vec{p}}{M} \\ \frac{\mathrm{d}\vec{p}}{\mathrm{d}t} &= \frac{\partial}{\partial\vec{p}} H(\vec{r},\vec{p}) = \frac{\partial U(\vec{p})}{\partial\vec{r}}. \end{aligned} \tag{1.23}$$

We notice that only the first dynamic equation (1.23) is in agreement with the first Hamilton equation (1.20), as the second equation is not in agreement with the second Hamilton equation (1.20) – a minus sign is missing. This sign is essential for the energy conservation:

$$\begin{aligned} \frac{\mathrm{d}E}{\mathrm{d}t} = \frac{\mathrm{d}H(\vec{r},\vec{p})}{\mathrm{d}t} &= \frac{\partial H(\vec{r},\vec{p})}{\partial\vec{r}} \frac{\mathrm{d}\vec{r}}{\mathrm{d}t} + \frac{\partial H(\vec{r},\vec{p})}{\partial\vec{p}} \frac{\mathrm{d}\vec{p}}{\mathrm{d}t} \\ &= \frac{\partial H(\vec{r},\vec{p})}{\partial\vec{r}} \frac{\partial H(\vec{r},\vec{p})}{\partial\vec{p}} - \frac{\partial H(\vec{r},\vec{p})}{\partial\vec{p}} \frac{\partial H(\vec{r},\vec{p})}{\partial\vec{r}} = 0. \end{aligned}$$

It is interesting to calculate the standard deviation

$$\Delta\vec{r} = \sqrt{\langle\vec{r}^2\rangle - \langle\vec{r}\rangle^2} \tag{1.24}$$

for a free particle obtained from the Schrödinger equation (1.9) with the classical Hamiltonian (1.8),

$$\frac{\partial}{\partial t}|\psi(\vec{r},t)\rangle = -\frac{\mathrm{i}}{\hbar}\frac{\vec{p}^2}{2M}|\psi(\vec{r},t)\rangle$$
$$\frac{\partial}{\partial t}\langle\psi(\vec{r},t)| = \frac{\mathrm{i}}{\hbar}\langle\psi(\vec{r},t)|\frac{\vec{p}^2}{2M}. \tag{1.25}$$

From these equations, for the time derivative of the first term of the deviation (1.24) we obtain

$$\begin{aligned}\frac{\partial}{\partial t}\langle\vec{r}^2\rangle &= \frac{\partial}{\partial t}\langle\psi(\vec{r},t)|\vec{r}^2|\psi(\vec{r},t)\rangle \\ &= \left(\frac{\partial}{\partial t}\langle\psi(\vec{r},t)|\right)\vec{r}^2|\psi(\vec{r},t)\rangle + \langle\psi(\vec{r},t)|\vec{r}^2\left(\frac{\partial}{\partial t}|\psi(\vec{r},t)\rangle\right). \\ &= \frac{\mathrm{i}}{2M\hbar}\langle\psi(\vec{r},t)|\left(\vec{p}^2\vec{r}^2 - \vec{r}^2\vec{p}^2\right)|\psi(\vec{r},t)\rangle.\end{aligned} \tag{1.26}$$

With the commutation formula (1.18), and the algebraic formulas for operator products,

$$[AB,C] = A[B,C] + [A,C]B$$
$$[A,BC] = B[A,C] + [A,B]C,$$

the operator of the expression in (1.24) takes a simpler form,

$$\begin{aligned}\frac{\mathrm{i}}{2M\hbar}\left(\vec{p}^2\vec{r}^2 - \vec{r}^2\vec{p}^2\right) &= -\frac{\mathrm{i}}{2M\hbar}\left[\vec{r}^2,\vec{p}^2\right] = -\frac{\mathrm{i}}{2M\hbar}\left\{\vec{r}\left[\vec{r},\vec{p}^2\right] + \left[\vec{r},\vec{p}^2\right]\vec{r}\right\} \\ &= -\frac{\mathrm{i}}{2M\hbar}\left\{\vec{r}\vec{p}[\vec{r},\vec{p}] + \vec{r}[\vec{r},\vec{p}]\vec{p} + \vec{p}[\vec{r},\vec{p}]\vec{r} + [\vec{r},\vec{p}]\vec{p}\vec{r}\right\} \\ &= \frac{3}{M}\left(\vec{r}\vec{p} + \vec{p}\vec{r}\right).\end{aligned} \tag{1.27}$$

At the same time, for the second tem of the standard deviation (1.24), from equations (1.25) we obtain

$$\begin{aligned}\frac{\partial}{\partial t}\langle\vec{r}\rangle &= \frac{\partial}{\partial t}\langle\psi(\vec{r},t)|\vec{r}|\psi(\vec{r},t)\rangle \\ &= \left(\frac{\partial}{\partial t}\langle\psi(\vec{r},t)|\right)\vec{r}|\psi(\vec{r},t)\rangle + \langle\psi(\vec{r},t)|\left(\vec{r}\frac{\partial}{\partial t}|\psi(\vec{r},t)\rangle\right) \\ &= \frac{\mathrm{i}}{2\hbar M}\langle\psi(\vec{r},t)|\left(\vec{p}^{2}\vec{r}-\vec{r}\vec{p}^{2}\right)|\psi(\vec{r},t)\rangle,\end{aligned} \tag{1.28}$$

depending on the operator

$$\frac{\mathrm{i}}{2\hbar M}\left(\vec{p}^{2}\vec{r}-\vec{r}\vec{p}^{2}\right)=\frac{\mathrm{i}}{2\hbar M}\left[\vec{p}^{2},\vec{r}\right]=\frac{\mathrm{i}}{2\hbar M}\left\{\vec{p}[\vec{p},\vec{r}]+[\vec{p},\vec{r}]\vec{p}\right\}=\frac{3\vec{p}}{M}. \tag{1.29}$$

With this expression, we obtain the time derivative (1.28),

$$\frac{\partial}{\partial t}\langle\vec{r}\rangle=3\frac{\langle\vec{p}\rangle}{M}, \tag{1.30}$$

which is identical with the first equation (1.19), as the time derivative of the second term of the standard deviation (1.24) is

$$\frac{\partial}{\partial t}\langle\vec{r}\rangle^{2}=2\langle\vec{r}\rangle\frac{\partial}{\partial t}\langle\vec{r}\rangle=\frac{6}{M}\langle\vec{r}\rangle\langle\vec{p}\rangle. \tag{1.31}$$

From this equation and equation (1.26) with (1.27), we obtain the time derivative of the standard deviation

$$\frac{\partial}{\partial t}\left(\langle\vec{r}^{2}\rangle-\langle\vec{r}\rangle^{2}\right)=\frac{3}{M}\left(\langle\vec{r}\vec{p}+\vec{p}\vec{r}\rangle-2\langle\vec{r}\rangle\langle\vec{p}\rangle\right). \tag{1.32}$$

In this equation, we consider Heisenberg time dependent operators, which, according to equation (1.30), satisfy the relation

$$3\vec{p}(t)=M\frac{\partial}{\partial t}\vec{r}(t)=M\dot{\vec{r}},$$

and coordinates as functions of velocities and time,

$$\vec{r} = \dot{\vec{r}}t = \frac{3\vec{p}}{M}t. \qquad \textbf{(1.33)}$$

With these expressions, equation (1.32) takes the form

$$\begin{aligned}\frac{\partial}{\partial t}\left(\langle\vec{r}^2\rangle - \langle\vec{r}\rangle^2\right) &= \frac{3}{M}\left(\langle\vec{r}\vec{p} + \vec{p}\vec{r}\rangle - 2\langle\vec{r}\rangle\langle\vec{p}\rangle\right) \\ &= \frac{3}{M}\left(\langle\dot{\vec{r}}\vec{p} + \vec{p}\dot{\vec{r}}\rangle - 2\langle\dot{\vec{r}}\rangle\langle\vec{p}\rangle\right)t \\ &= \frac{9\cdot 2}{M^2}\left(\langle\vec{p}^2\rangle - \langle\vec{p}\rangle^2\right)t.\end{aligned}$$

By integration with time, we obtain the square of the time deviation (1.24)

$$\begin{aligned}\Delta\vec{r}^2 = \langle\vec{r}^2\rangle - \langle\vec{r}\rangle^2 &= \frac{9}{M^2}\left(\langle\vec{p}^2\rangle - \langle\vec{p}\rangle^2\right)t^2 \\ &= \frac{9}{M^2}\Delta\vec{p}^2 t^2,\end{aligned}$$

as a function of the momentum standard deviation

$$\Delta\vec{p} = \sqrt{\left(\langle\vec{p}^2\rangle - \langle\vec{p}\rangle^2\right)}. \qquad \textbf{(1.34)}$$

Thus, for a free quantum particle, from the Schrödinger equation (1.9) we obtain a coordinate deviation continuously increasing with time:

$$\Delta\vec{r} = 3\frac{\Delta\vec{p}}{M}t. \qquad \textbf{(1.35)}$$

This weird consequence of the Schrödinger equation (1.9), describing a free particle by a distribution rapidly spreading in space, the also weird factor 3 of the first equation (1.19), and the solution (1.21) with (1.22) of this equation, describing the dynamics (1.23) contradictory to the Hamilton equations (1.20), suggests a reconsideration of the quantum dynamics.

This reconsideration is based on Einstein's fundamental equations regarding the energy $E = \vec{p}^2 / 2M$ and the momentum $\vec{p}$ of a free quantum particle as functions of their wavy characteristics, the frequency ω and the wave vector $\vec{k}$:

$$\begin{aligned} E &= \frac{\vec{p}^2}{2M} = \hbar\omega \\ \vec{p} &= \hbar\vec{k}. \end{aligned} \tag{1.36}$$

Such a particle is described by two wave functions, in the coordinate and the momentum conjugate spaces:

$$\begin{aligned} \psi(\vec{r},t) &= \frac{1}{(2\pi\hbar)^{3/2}} \int \varphi(\vec{p},t) e^{\frac{i}{\hbar}\left[\vec{p}\vec{r} - \frac{\vec{p}^2}{2M}t\right]} \mathrm{d}^3\vec{p} \\ \varphi(\vec{p},t) &= \frac{1}{(2\pi\hbar)^{3/2}} \int \psi(\vec{r},t) e^{-\frac{i}{\hbar}\left[\vec{p}\vec{r} - \frac{\vec{p}^2}{2M}t\right]} \mathrm{d}^3\vec{r}, \end{aligned} \tag{1.37}$$

which, with (1.10) and (1.11), satisfy the two fundamental equations (1.36):

$$\begin{aligned} H\psi(\vec{r},t) &= i\hbar\frac{\partial}{\partial t}\psi(\vec{r},t) = \frac{1}{(2\pi\hbar)^{3/2}} i\hbar\frac{\partial}{\partial t}\int \varphi(\vec{p},t) e^{\frac{i}{\hbar}\left[\vec{p}\vec{r} - \frac{\vec{p}^2}{2M}t\right]} \mathrm{d}^3\vec{p} = \frac{\vec{p}^2}{2M}\psi(\vec{r},t) \\ \vec{p}\varphi(\vec{p},t) &= -i\hbar\frac{\partial}{\partial \vec{r}}\varphi(\vec{p},t) = -\frac{1}{(2\pi\hbar)^{3/2}}\frac{\partial}{\partial \vec{r}}\int \psi(\vec{r},t) e^{-\frac{i}{\hbar}\left[\vec{p}\vec{r} - \frac{\vec{p}^2}{2M}t\right]} \mathrm{d}^3\vec{r} = \vec{p}\varphi(\vec{p},t). \end{aligned} \tag{1.38}$$

It is remarkable that the dynamics of such a particle is described by an invariant distribution function, propagating with a velocity $\dot{\vec{r}}$ equal to the wave velocity:

$$\begin{aligned} \frac{\partial}{\partial \vec{p}}\frac{\vec{p}^2}{2M} &= \frac{\vec{p}}{M} = \dot{\vec{r}} \\ \frac{\partial}{\partial \vec{r}}\frac{\vec{p}^2}{2M} &= 0. \end{aligned} \tag{1.39}$$

Thus, we notice that by the reduction of the two equations (1.38) to only one, according to the Schrödinger equation (1.9) with the Hamiltonian (1.8),

$$\mathrm{i}\hbar\frac{\partial}{\partial t}\psi(\vec{r},t)=\frac{\vec{p}^2}{2M}\psi(\vec{r},t), \tag{1.40}$$

with the solution

$$\psi(\vec{r},t)=\frac{1}{(2\pi\hbar)^{3/2}}\int\varphi(\vec{p})e^{-\frac{\mathrm{i}}{\hbar}\frac{\vec{p}^2}{2M}t}\mathrm{d}^3\vec{p}, \tag{1.41}$$

the invariance distribution property of the wave functions (1.38) is broken, leading to a rapidly spreading solution according to equation (1.35). The generalization of equation (1.40) for a potential $U(\vec{r})$ [3], which is the Schrödinger equation (1.9) with the Hamiltonian (1.8), leads to the first dynamic equation (1.19), which violates the firs Hamilton equation (1.20) with a factor 3. More than that, if for the Schrödinger equation, we consider a solution with a propagation factor, of the form (1.21)-(1.22), we obtain the dynamic equations (1.23), in agreement with the first equation (1.20), but contradictory to the second equation (1.20). These difficulties can be avoided by generalizing the fundamental wave functions (1.37) for a potential $U(\vec{r})$, not under the form (1.21)-(1.22) which leads to the Schrödinger equation (1.9) with the Hamiltonian (1.8), but under the form

$$\begin{aligned}\psi(\vec{r},t)&=\frac{1}{(2\pi\hbar)^{3/2}}\int\varphi(\vec{p},t)e^{\frac{\mathrm{i}}{\hbar}\{\vec{p}\vec{r}-[T(\vec{p})-U(\vec{r})]t\}}\mathrm{d}^3\vec{p}\\&=\frac{1}{(2\pi\hbar)^{3/2}}\int\varphi(\vec{p},t)e^{\frac{\mathrm{i}}{\hbar}[\vec{p}\vec{r}-L(\vec{p},\vec{r})t]}\mathrm{d}^3\vec{p}\\\varphi(\vec{p},t)&=\frac{1}{(2\pi\hbar)^{3/2}}\int\psi(\vec{r},t)e^{-\frac{\mathrm{i}}{\hbar}\{\vec{p}\vec{r}-[T(\vec{p})-U(\vec{r})]t\}}\mathrm{d}^3\vec{r}\\&=\frac{1}{(2\pi\hbar)^{3/2}}\int\psi(\vec{r},t)e^{-\frac{\mathrm{i}}{\hbar}[\vec{p}\vec{r}-L(\vec{p},\vec{r})t]}\mathrm{d}^3\vec{r},\end{aligned} \tag{1.42}$$

where we replaced the Hamiltonian (1.8) in (1.21) and (1.22) by the Lagrangian

$$\begin{aligned} L(\vec{p},\vec{r}) &= \vec{p}\dot{\vec{r}} - H(\vec{p},\vec{r}) = \frac{\vec{p}^2}{M} - \left(\frac{\vec{p}^2}{2M} + U(\vec{r})\right) \\ &= \frac{\vec{p}^2}{2M} - U(\vec{r}) = T(\vec{p}) - U(\vec{r}) \end{aligned} \quad \textbf{(1.43)}$$

The group velocities of these equations,

$$\begin{aligned} \frac{\mathrm{d}\vec{r}}{\mathrm{d}t} &= \frac{\partial T(\vec{p})}{\partial \vec{p}} = \frac{\partial}{\partial \vec{p}} H(\vec{p},\vec{r}) \\ \frac{\mathrm{d}\vec{p}}{\mathrm{d}t} &= -\frac{\partial U(\vec{r})}{\partial \vec{r}} = -\frac{\partial}{\partial \vec{r}} H(\vec{p},\vec{r}), \end{aligned} \quad \textbf{(1.44)}$$

are in agreement with the Hamilton equations (1.20). According to equations (1.39), the first equation (1.44) describes an invariant distribution of matter propagating in time. At the same time, the wave functions (1.42) suggest the consideration of the relativistic Lagrangian

$$L(x^\alpha, v^\alpha) = -Mc^2\sqrt{g_{\alpha\beta}v^\alpha v^\beta} = -Mc^2, \quad \textbf{(1.45)}$$

as an invariant depending on the metric tensor $g_{\alpha\beta}$ and the velocities the proper time τ,

$$v^\alpha = \frac{\mathrm{d}x^\alpha}{\mathrm{d}s} = \frac{\mathrm{d}x^\alpha}{c\mathrm{d}\tau}. \quad \textbf{(1.46)}$$

In agreement with the relativistic invariance of the time-space interval,

$$\mathrm{d}s \equiv c\mathrm{d}\tau = \sqrt{g_{\alpha\beta}\mathrm{d}x^\alpha \mathrm{d}x^\beta}, \quad \textbf{(1.47)}$$

we obtain the fundamental equation

$$g_{\alpha\beta}v^\alpha v^\beta = 1. \quad \textbf{(1.48)}$$

With this Lagrangian, the wave functions (1.42) in the proper time, are of the form

$$\psi\left(x^{i},\tau\right)=\frac{1}{\left(2\pi\hbar\right)^{3/2}}\int\varphi\left(P^{j},\tau\right)e^{\frac{i}{\hbar}\left[P^{j}x^{j}-L\left(x^{\alpha},v^{\alpha}\right)\tau\right]}\mathrm{d}^{3}P$$

$$\varphi\left(P^{j},\tau\right)=\frac{1}{\left(2\pi\hbar\right)^{3/2}}\int\psi\left(x^{i},\tau\right)e^{-\frac{i}{\hbar}\left[P^{j}x^{j}-L\left(x^{\alpha},v^{\alpha}\right)\tau\right]}\mathrm{d}^{3}x. \tag{1.49}$$

In this case, the distribution in the coordinate space x^{j} of the wave function (1.50) is determined only by the momentum spectrum $\varphi\left(P^{j},\tau\right)$ and the propagation factor with the coordinate dependent phase $P^{j}x^{j}$, as an invariant distribution, as this spectrum is constant, $\varphi\left(P^{j},\tau\right)=\varphi\left(P^{j}\right)$. In this framework, the theory of relativity [1, 2], and quantum mechanics [3], form a unitary domain of physics.

In chapter 2, we describe a quantum particle as an invariant distribution of matter, represented by a Fourier series expansion as a wave function with the time dependent phase proportional to the relativistic Lagrangian. We obtain a quantization relation from the mass in this Lagrangian with the total mass as an integral of the matter density. In chapter 3, we present the general theory of relativity in Dirac's formalism, and the dynamics of a quantum particle in a gravitational system with spherical symmetry. In chapter 4, we consider a quantum particle in electromagnetic field, and derive the particle-field dynamics in a flat space. In section 5, from the total action of the gravitational field, matter, electromagnetic field, and electric charge, we obtain Einstein's gravitation law, Lorentz's force, and generalized Maxwell equations in the framework of the general theory of relativity.

CONCLUSION

We obtained a unitary framework of the general theory of relativity, describing a physical system in a four-dimensional system of time-space coordinates with an invariant time-space interval, and of quantum mechanics, with wave functions depending the relativistic Lagrangian in the time dependent phases, which describe invariant distributions propagating in space.

CHAPTER 2

Quantum Particle as a Distribution of Matter

Abstract: In this chapter, we consider the wave function of a particle as a wave-packet describing the density amplitude in the coordinate space, and its inverse Fourier transform, as a wave function in the momentum space. We obtain the momentum from the mass Lagrangian in the time-dependent phase of the wave function, and the particle dynamics from the group velocities of these wave-packets in the two conjugate spaces of the coordinates and of the momentum. From the equality of the mass in the relativistic Lagrangian, which describes the matter dynamics, with the total mass as an integral of the density, we obtain the matter quantization.

Keywords: Wave function, Wave packet, Group velocity, Canonical momentum, Proper time, Lagrangian, Amplitude function, Distribution function, Normalization, Mass density, Metric tensor, Quantization rule.

We consider the wave functions (1.42) with the Lagrangian (1.45), in the proper time $t=\tau$,

$$\begin{aligned}\psi\left(x^i,\tau\right)&=\frac{1}{\left(2\pi\hbar\right)^{3/2}}\int\varphi\left(P^j,\tau\right)e^{\frac{\mathrm{i}}{\hbar}\left[P^j x^j+Mc^2\sqrt{g_{\alpha\beta}v^\alpha v^\beta}\,\tau\right]}\mathrm{d}^3P\\ \varphi\left(P^j,\tau\right)&=\frac{1}{\left(2\pi\hbar\right)^{3/2}}\int\psi\left(x^i,\tau\right)e^{-\frac{\mathrm{i}}{\hbar}\left[P^j x^j+Mc^2\sqrt{g_{\alpha\beta}v^\alpha v^\beta}\,\tau\right]}\mathrm{d}^3x.\end{aligned} \quad (2.1)$$

with the normalization conditions

$$\int\left|\psi\left(x^i,\tau\right)\right|^2\mathrm{d}^3x=1,\qquad\int\left|\varphi\left(P^i,\tau\right)\right|^2\mathrm{d}^3P=1, \quad (2.2)$$

for the distribution functions $\left|\psi\left(x^i,\tau\right)\right|^2$ and $\left|\varphi\left(P^i,\tau\right)\right|^2$ in the two conjugate spaces. We notice that with the invariant Lagrangian (1.45) in the time-dependent phase,

Eliade Stefanescu

the first equation (2.1) describes an invariant distribution function $\left|\psi\left(x^i,\tau\right)\right|^2$ propagating in the coordinate space, as the second equation (2.1), for a free particle describes a constant distribution function $\left|\varphi\left(P^i,\tau\right)\right|^2=\left|\varphi\left(P^i\right)\right|^2$ in the conjugate space of the momentum. This suggests the interpretation of the wave function $\psi\left(x^i,\tau\right)$ as the amplitude function of a matter distribution with the mass density

$$\rho\left(x^i,\tau\right)=M_0\left|\psi\left(x^i,\tau\right)\right|^2 \qquad \textbf{(2.3)}$$

These wave functions, with the Lagrangian (1.45),

$$L\left(x^\alpha,v^\alpha\right)=-Mc^2\sqrt{g_{\alpha\beta}v^\alpha v^\beta}\ , \qquad \textbf{(2.4)}$$

depend on the canonical momentum

$$\begin{aligned} P^j &= \frac{\partial L}{c\partial v^j}=-Mc^2\frac{\partial}{c\partial v^j}\sqrt{g_{\alpha\beta}v^\alpha v^\beta} \\ &= -Mc\frac{1}{2\sqrt{g_{\alpha\beta}v^\alpha v^\beta}}\frac{\partial}{\partial v^j}\left(g_{j\beta}v^j v^\beta+g_{\alpha j}v^\alpha v^j\right) \\ &= -Mc\frac{g_{j\beta}v^\beta+g_{\alpha j}v^\alpha}{2\sqrt{g_{\alpha\beta}v^\alpha v^\beta}}=-Mcg_{j\mu}v^\mu , \end{aligned} \qquad \textbf{(2.5)}$$

where we used and the fundamental equation (1.48). The dynamics of these distributions of matter are described by the two group velocities

$$\begin{aligned}
\frac{\mathrm{d}}{\mathrm{d}\tau}x^j &= \frac{\partial L}{\partial P^j} = \frac{\partial\left(-Mc^2\sqrt{g_{\alpha\beta}v^\alpha v^\beta}\right)}{\partial\left(-Mcg_{j\mu}v^\mu\right)} \\
&= \frac{c}{2\sqrt{g_{\alpha\beta}v^\alpha v^\beta}}\frac{\partial}{\partial\left(g_{j\mu}v^\mu\right)}\left(g_{j\beta}v^j v^\beta + g_{\alpha j}v^\alpha v^j\right) \\
&= c\frac{2v^j}{2\sqrt{g_{\alpha\beta}v^\alpha v^\beta}} = cv^j = \dot{x}^j \\
\frac{\mathrm{d}}{\mathrm{d}\tau}P^j &= \frac{\partial L}{\partial x^j} = \frac{\partial\left(-Mc^2\sqrt{g_{\alpha\beta}v^\alpha v^\beta}\right)}{\partial x^j} \\
&= \frac{-Mc^2 v^\alpha v^\beta}{2\sqrt{g_{\alpha\beta}v^\alpha v^\beta}}\frac{\partial}{\partial x^j}g_{\alpha\beta} = -\frac{1}{2}Mc^2 v^\alpha v^\beta g_{\alpha\beta,j}\,.
\end{aligned} \tag{2.6}$$

In a flat space, with the metric tensor elements $g_{jj} = -1$, the momentum (2.5) takes the classical form

$$P^j = Mcv^j = Mc\frac{\mathrm{d}x^j}{\mathrm{d}s} = Mc\frac{\mathrm{d}x^j}{c\mathrm{d}\tau} = M\frac{\mathrm{d}x^j}{\mathrm{d}\tau},$$

of the product of the mass M with the velocity. From the first equation (2.6), we notice that the velocity $\frac{\mathrm{d}}{\mathrm{d}\tau}x^j = \frac{\partial L}{\partial P^j}$ of the waves describing the density propagation is equal to the matter velocity $\dot{x}^j = cv^j$. We also notice that the second equation (2.6) represents a force proportional to the mass M and the gradient of the metric tensor as a potential. From this perspective, we consider the equality of the mass M_0 obtained by the integration of the matter density (2.3), with the mass M describing the dynamics of this matter by the Lagrangian (2.4), as a quantization rule:

$$\int\rho\left(x^i,\tau\right)\mathrm{d}^3x = M_0\int\left|\psi\left(x^i,\tau\right)\right|^2\mathrm{d}^3x = M_0 = M\,. \tag{2.7}$$

CONCLUSION

We considered a quantum particle as a distribution of matter in motion according to the general theory of relativity. From the Fourier series expansions of the density amplitude, in the coordinate and the momentum spaces, we obtained the particle dynamics as the group velocities of the wave-packets propagating in these spaces, which are of the form of the Hamilton equations. In this representation, the matter distribution in the momentum space determines the matter distribution in the coordinate space, and conversely, the matter distribution in the coordinate space determines the matter distribution in the momentum space. Since the relativistic Lagrangian is invariant, the matter distributions in the coordinate and momentum spaces are invariant, propagating with the group velocities in these spaces. From the total mass as the integral of the matter density, and the mass in the Lagrangian of the time-dependent phase, which describes the dynamics of this density, we obtained the quantization condition as the equality of the two masses.

CHAPTER 3

Quantum Particle in the Gravitational Field

Abstract: In this chapter, we demonstrate the main theorems of the general theory of relativity, and derive dynamic equations for a quantum particle in a gravitational field. For two curvilinear time-space systems of reference, we obtain the transformed matrices of vectors and tensors. We define contravariant and covariant representations, and the quotient theorem. We consider the curvature of the physical four-dimensional time-space system, in the total system, including a number of extra-coordinates, enabling this curvature. For the physical system, as a hypersurface in the total system, we obtain the metric tensor, as a function of the extra-coordinates, which, on this hypersurface, depend on the physical coordinates, as parameters. In this framework, we obtain the Christoffel symbols as functions of the coordinate derivatives of the metric elements. We obtain the covariant derivative, including only physical effects, without the curvature effects described by the ordinary derivative. We show that the covariant derivative of a metric element is null, and derive equations for the matter conservation, harmonic oscillations, and curvilinear forms of the Gauss and Stokes theorems. From the inertial and gravitational forces, we obtain the geodesic equations of a null covariant acceleration, as from the invariance of the time-space interval, we obtain that any vector is perpendicular to its covariant derivative with any coordinate. We define the curvature tensor and derive the symmetry and the Bianci relations. We define the Ricci tensor, and from the Bianci relation for this tensor we obtain Einstein's law of gravitation. From this law for a gravitational system with spherical symmetry, we obtain the Schwarzschild solution for the metric tensor. We derive the Einstein gravitational law in the presence of matter, and study the dynamics of a quantum particle in a black hole. Outside a black hole we consider a time-like region, with a far region where all the bodies are attracted, and a near region, where the coming bodies are strongly decelerated, to a null velocity at the Schwarzschild boundary, but reaching this boundary only in an infinite time. The internal part of a black hole is defined as a space-like region, where at its formation, the central matter explodes, having the tendency to concentrate at the Schwarzschild boundary, but reaching this boundary also in an infinite time.

Keywords: Contravariant vector, Covariant vector, Tensor, Index contraction, Dummy index, Index raising, Index lowering, Curvature, Curved time-space coordinates, Curved physical system, Physical universe, Total universe, Physical hypersurface, Metric tensor, Quotient theorem, Christoffel symbol, Covariant derivative, Scalar density, Tensor density, Conservation equation, Harmonic oscillation, D'Alembert equation, Laplace equation, Gauss theorem, Stokes theorem, Geodesic, Curvature tensor, Ricci tensor, Bianci relations, Newton's

Eliade Stefanescu

gravitation law, Einstein's gravitation law, Redshift, Schwarzschild solution, Schwarzschild radius, Black hole, Big Bang, Inflation.

3.1. CURVED TIME-SPACE COORDINATES

The physical reality we live in is described in a four-dimensional system of time-space coordinates, with the invariance property of the time-space interval (1.47),

$$ds^2 \equiv c^2 d\tau^2 = g_{\alpha\beta} dx^\alpha dx^\beta = g_{\alpha'\beta'} dx^{\alpha'} dx^{\beta'} . \qquad \textbf{(3.1)}$$

For a physical system, we consider the transformation from a system of coordinates $S\left(x^\alpha\right)$ to another system of coordinates $S'\left(x^{\alpha'}\right)$,

$$dx^\alpha = \frac{\partial x^\alpha}{\partial x^{\alpha'}} dx^{\alpha'} \doteq x^\alpha{}_{,\alpha'} dx^{\alpha'} , \qquad \textbf{(3.2)}$$

with the inverse transformation

$$dx^{\alpha'} = \frac{\partial x^{\alpha'}}{\partial x^\alpha} dx^\alpha \doteq x^{\alpha'}{}_{,\alpha} dx^\alpha . \qquad \textbf{(3.3)}$$

These coordinates essentially describe the symmetry of the system. For example, in a system with spherical symmetry, we consider the spherical coordinates $x^1 = r$, $x^2 = \theta$, $x^3 = \phi$, as the time-space interval (3.1) is of the form

$$ds^2 = dx^{0^2} - dr^2 - r^2\left(d\theta^2 + \sin^2\theta d\phi^2\right), \qquad \textbf{(3.4)}$$

with the metric elements

$$g_{00} = 1, \qquad g_{11} = -1, \qquad g_{22} = -r^2, \qquad g_{33} = -r^2 \sin^2\theta . \qquad \textbf{(3.5)}$$

Any four-vector A^μ satisfying the transformation relations (3.2)-(3.3) is called four-vector of the physical system. With these transformation equations,

$$\begin{aligned} A^\alpha &= x^\alpha{}_{,\alpha'} A^{\alpha'} \\ A^{\alpha'} &= x^{\alpha'}{}_{,\alpha} A^\alpha , \end{aligned} \qquad \textbf{(3.6)}$$

we obtain the identity

$$A^{\alpha} = x^{\alpha}{}_{,\alpha'} A^{\alpha'} = x^{\alpha}{}_{,\alpha'} x^{\alpha'}{}_{,\beta} A^{\beta} = \delta^{\alpha}_{\beta} A^{\beta},$$

where δ^{α}_{β} is the unit matrix, as the product of the two transformation matrices are inverse to one another:

$$x^{\alpha}{}_{,\alpha'} x^{\alpha'}{}_{,\beta} = \delta^{\alpha}_{\beta}. \quad \textbf{(3.7)}$$

Considering an equation similar to (3.1), with the transformation relations (3.6), for the square of the amplitude of the four-vector A^{μ},

$$\begin{aligned} |A|^2 &= (A, A) = g_{\alpha\beta} A^{\alpha} A^{\beta} \\ &= g_{\alpha'\beta'} A^{\alpha'} A^{\beta'} = g_{\alpha'\beta'} x^{\alpha'}{}_{,\alpha} x^{\beta'}{}_{\beta} A^{\alpha} A^{\beta}, \end{aligned} \quad \textbf{(3.8)}$$

we obtain the transformation relation

$$g_{\alpha\beta} = x^{\alpha'}{}_{,\alpha} x^{\beta'}{}_{\beta} g_{\alpha'\beta'}. \quad \textbf{(3.9)}$$

Any two index quantity $T_{\mu\nu}$ satisfying such a transformation,

$$T_{\alpha\beta} = x^{\alpha'}{}_{,\alpha} x^{\beta'}{}_{\beta} T_{\alpha'\beta'}, \quad \textbf{(3.10)}$$

is called tensor, as $g_{\alpha\beta}$ is called the metric tensor. By an index interchange in (3.1),

$$\mathrm{d}s^2 = g_{\alpha\beta} \mathrm{d}x^{\alpha} \mathrm{d}x^{\beta} = g_{\beta\alpha} \mathrm{d}x^{\beta} \mathrm{d}x^{\alpha} \equiv g_{\beta\alpha} \mathrm{d}x^{\alpha} \mathrm{d}x^{\beta},$$

we find that the metric tensor is symmetric:

$$g_{\alpha\beta} = g_{\beta\alpha}. \quad \textbf{(3.11)}$$

For the four-vector $A = (A^{\mu})$, we consider the squared amplitude (3.8) of the form

$$|A|^2 = (A, A) = g_{\alpha\beta} A^{\alpha} A^{\beta} = A_{\alpha} A^{\alpha} = A_{\alpha'} A^{\alpha'}, \quad \textbf{(3.12)}$$

depending on the components A^{α} called contravariant components, and the components

$$A_{\alpha} = g_{\alpha\beta} A^{\beta}, \tag{3.13}$$

called covariant components, as the invariant quantity $|A|^2 = (A, A)$ is a scalar. According to (3.13), the metric tensor $g_{\alpha\beta}$ is called the lowering tensor. Similarly, we define the raising tensor $g^{\alpha\beta}$,

$$A^{\alpha} = g^{\alpha\beta} A_{\beta}. \tag{3.14}$$

Multiplying the two equations,

$$A_{\alpha} A^{\alpha} = g_{\alpha\beta} g^{\alpha\gamma} A^{\beta} A_{\gamma} = \delta_{\beta}^{\gamma} A_{\gamma} A^{\beta},$$

with the symmetry relation (3.11), we find that the two matrices representing the lowering and the raising tensors are inverse to one another:

$$g_{\beta\alpha} g^{\alpha\gamma} = \delta_{\beta}^{\gamma}. \tag{3.15}$$

With this relation, from (3.5) for a physical system with spherical symmetry, we obtain the elements of the raising tensor:

$$g^{00} = 1, \qquad g^{11} = -1, \qquad g^{22} = -r^{-2}, \qquad g^{33} = -r^{-2} \sin^{-2} \theta. \tag{3.16}$$

The raising tensor satisfies a transformation relation similar to (3.9),

$$g^{\alpha\beta} = x^{\alpha}{}_{,\alpha'} x^{\beta}{}_{,\beta'} g^{\alpha'\beta'}, \tag{3.17}$$

as by multiplying this equation with (3.9), from (3.7), we reobtain (3.15):

$$\begin{aligned} g^{\alpha\beta} g_{\beta\gamma} &= x^{\alpha}{}_{,\alpha'} x^{\beta}{}_{,\beta'} x^{\beta''}{}_{,\beta} x^{\gamma''}{}_{,\gamma} g^{\alpha'\beta'} g_{\beta''\gamma''} = x^{\alpha}{}_{,\alpha'} \delta_{\beta'}^{\beta''} x^{\gamma''}{}_{,\gamma} g^{\alpha'\beta'} g_{\beta''\gamma''} \\ &= x^{\alpha}{}_{,\alpha'} x^{\gamma''}{}_{,\gamma} g^{\alpha'\beta'} g_{\beta'\gamma''} = x^{\alpha}{}_{,\alpha'} x^{\gamma''}{}_{,\gamma} \delta_{\gamma''}^{\alpha'} = x^{\alpha}{}_{,\alpha'} x^{\alpha'}{}_{,\gamma} \\ &= \delta_{\gamma}^{\alpha}. \end{aligned}$$

With the transformation relations (3.6) and (3.9), for the covariant four-vector (3.13) we obtain a transformation similar to the transformation (3.6) for a contravariant four-vector,

$$A_\alpha = x^{\alpha'}{}_{,\alpha} x^{\beta'}{}_{,\beta} g_{\alpha'\beta'} x^{\beta}{}_{,\beta'} A^{\beta'} = x^{\alpha'}{}_{,\alpha} A_{\alpha'} . \quad \textbf{(3.18)}$$

We also notice that with the transformation relations (3.6) and (3.18) for a contravariant and a covariant four-vector, the transformation relation (3.10) can be generalized for a tensor with arbitrary numbers of contravariant and covariant indices:

$$T^{\alpha}{}_{\beta\gamma}{}^{\delta \ldots}{}_{\ldots} = x^{\alpha}{}_{,\alpha'} x^{\beta'}{}_{,\beta} x^{\gamma'}{}_{,\gamma} x^{\delta}{}_{,\delta'} T^{\alpha'}{}_{\beta'\gamma'}{}^{\delta' \ldots}{}_{\ldots} \quad \textbf{(3.19)}$$

With the contravariant and the covariant components of the four-vectors A and B we define the scalar product of these four-vectors:

$$(A, B) = A_\mu B^\mu = g_{\mu\alpha} A^\alpha g^{\mu\beta} B_\beta = \delta_\alpha^\beta A^\alpha B_\beta = B_\alpha A^\alpha = (B, A) \quad \textbf{(3.20)}$$

This means that the scalar product is symmetric, and that the positions up and down of a dummy index can be interchanged. We can show that the scalar product (3.20) is invariant. For this, we consider the squared amplitude of a linear combination of the two vectors with an arbitrary parameter λ,

$$\begin{aligned} |A + \lambda B|^2 &= (A + \lambda B, A + \lambda B) = (A, A) + \lambda^2 (B, B) + 2\lambda (A, B) \\ &= |A|^2 + \lambda^2 |B|^2 + 2\lambda (A, B). \end{aligned} \quad \textbf{(3.21)}$$

From the invariance of the amplitudes for an arbitrary parameter λ, we obtain that the scalar product (A, B) is invariant. In a matrix representation, the scalar product is of the form:

$$(A,B) = A_\mu B^\mu = \begin{pmatrix} A_0 & A_1 & A_2 & A_3 \end{pmatrix} \begin{pmatrix} B^0 \\ B^1 \\ B^2 \\ B^3 \end{pmatrix} = A^\mu B_\mu$$

$$= B_\mu A^\mu = \begin{pmatrix} B_0 & B_1 & B_2 & B_3 \end{pmatrix} \begin{pmatrix} A^0 \\ A^1 \\ A^2 \\ A^3 \end{pmatrix} = (B,A). \quad \textbf{(3.22)}$$

<u>The quotient theorem</u>

From (3.7), we obtain the quotient theorem, which, in the simpler case considered here means that if the product of the quantity $Q_{\alpha\beta}$ with a four-vector A^α is a four-vector B_β, this quantity is a tensor. With the transformation relations (3.6) and (3.18) for the two vectors,

$$Q_{\alpha\beta} A^\alpha = Q_{\alpha\beta} x^\alpha{}_{,\alpha'} \underline{A^{\alpha'}}$$
$$= B_\beta = x^{\beta'}{}_{,\beta} B_{\beta'} = x^{\beta'}{}_{,\beta} Q_{\alpha'\beta'} \underline{A^{\alpha'}},$$

we obtain

$$Q_{\alpha\beta} x^\alpha{}_{,\alpha'} = x^{\beta'}{}_{,\beta} Q_{\alpha'\beta'}.$$

By multiplying this relation with $x^{\alpha'}{}_{\gamma}$,

$$Q_{\alpha\beta} x^\alpha{}_{,\alpha'} x^{\alpha'}{}_{\gamma} = Q_{\alpha\beta} \delta^\alpha_\gamma = Q_{\gamma\beta} = x^{\alpha'}{}_{\gamma} x^{\beta'}{}_{,\beta} Q_{\alpha'\beta'},$$

we obtain the tensorial transformation of the form,

$$Q_{\alpha\beta} = x^{\alpha'}{}_{\alpha} x^{\beta'}{}_{,\beta} Q_{\alpha'\beta'}, \quad \textbf{(3.23)}$$

which means that the two index quantity $Q_{\alpha\beta}$ is a tensor. Evidently, this demonstration can be generalized for an arbitrary tensor, with up and down indices.

3.2. CURVED PHYSICAL SYSTEM IN THE TOTAL UNIVERSE

According to Einstein's theory of relativity, the matter dynamics in a gravitational field is based on the equivalence of this field with the acceleration in this field: a body falling in a gravitational field does not feel this field anymore [1]. This means that the gravitation is described by a curvature of the coordinates, which describes the curvature of the physical system. A curved 4-dimensional physical system can be conceived only in a total universe, with a larger number of dimensions N where this system is curved [2]. We consider a flat total universe with the coordinates z^n, $n=1, 2,.....,N$, and a distance between two neighboring points

$$\mathrm{d}s^2 = h_{nm}\mathrm{d}z^n\mathrm{d}z^m, \tag{3.24}$$

depending on constant metric elements $h_{nm} = h_{mn}$, and the contravariant coordinate differentials $\mathrm{d}z^n$. With the covariant components

$$\mathrm{d}z_n = h_{nm}\mathrm{d}z^m, \tag{3.25}$$

the invariant differential interval (3.24) is of the form

$$\mathrm{d}s^2 = \mathrm{d}z_n\mathrm{d}z^n. \tag{3.26}$$

In this system with N coordinates z^n, our physical system with the 4 physical coordinates

$$\left(x^\mu\right)=\left(x^0,x^1,x^2,x^3\right)$$

is a four-dimensional hypersurface with four physical coordinates as parameters:

$$\begin{cases} z^1 = z^1\left(x^0,x^1,x^2,x^3\right) \\ z^2 = z^2\left(x^0,x^1,x^2,x^3\right) \\ \vdots \\ z^N = z^N\left(x^0,x^1,x^2,x^3\right). \end{cases} \tag{3.27}$$

By eliminating the last 4 universal coordinates as functions of the physical coordinates,

$$\begin{cases} z^{N-3} = z^{N-3}\left(x^0, x^1, x^2, x^3\right) \\ z^{N-2} = z^{N-2}\left(x^0, x^1, x^2, x^3\right) \\ z^{N-1} = z^{N-1}\left(x^0, x^1, x^2, x^3\right) \\ z^{N} = z^{N}\left(x^0, x^1, x^2, x^3\right), \end{cases}$$

we obtain the physical hypersurface described by $N-4$ equations depending on the physical coordinates as parameters:

$$\begin{cases} z^1 = z^1\left(z^{N-3}, z^{N-2}, z^{N-1}, z^N\right) = z^1\left(x^0, x^1, x^2, x^3\right) \\ z^2 = z^2\left(z^{N-3}, z^{N-2}, z^{N-1}, z^N\right) = z^2\left(x^0, x^1, x^2, x^3\right) \\ \vdots \\ z^{N-4} = z^{N-4}\left(z^{N-3}, z^{N-2}, z^{N-1}, z^N\right) = z^{N-4}\left(x^0, x^1, x^2, x^3\right). \end{cases} \quad \textbf{(3.28)}$$

From these equations we obtain the differential of the coordinate vector of the total universe,

$$\mathrm{d}z^n = z^n{}_{,\alpha}\mathrm{d}x^\alpha, \qquad n = 1, 2, \ldots\ N-4, \quad \textbf{(3.29)}$$

as a vector of the physical hypersurface. This means that any vector of the total universe of the form

$$A^n = z^n{}_{,\alpha}A^\alpha \quad \textbf{(3.30)}$$

is a vector of the physical hypersurface. Since the metric elements h_{nm} of the total universe are constant, we can define the covariant vector

$$z_{n,\alpha} = h_{nm}z^m{}_{,\alpha}, \quad \textbf{(3.31)}$$

and the differential of the covariant coordinate vector

$$\mathrm{d}z_n = z_{n,\alpha}\mathrm{d}x^\alpha, \quad \textbf{(3.32)}$$

as a vector of the physical hypersurface in the total universe has the covariant components

$$A_n = z_{n,\alpha} A^\alpha . \tag{3.33}$$

With these expressions, the differential interval (3.24) takes a form depending only on the physical coordinates

$$\begin{aligned} ds^2 &= h_{nm} z^n{}_{,\alpha} z^m{}_{,\beta} dx^\alpha dx^\beta, \qquad n,m = 1, 2,\ldots.., N-4 \\ &= g_{\alpha\beta} dx^\alpha dx^\beta, \end{aligned} \tag{3.34}$$

with the metric tensor

$$g_{\alpha\beta} = h_{nm} z^n{}_{,\alpha} z^m{}_{,\beta} = z_{n,\alpha} z^n{}_{,\beta} \tag{3.35}$$

We notice that in a curved system of coordinates, a coordinate derivative of a physical vector $A^\alpha{}_{,\beta}$ includes two terms: (1) a term due to the coordinate curvature, as a variation for a parallel displacement of the vector $A^n(x+dx) = A^n(x) = z^n{}_{,\beta} A^\beta(x)$ in the universal system, and (2) a term for the vector variation due to some external fields, not described by the coordinates. By a parallel displacement, besides the tangential component $A_{\tan}{}^n(x+dx)$ contained in the physical hypersurface, this vector gets a perpendicular component to the physical hypersurface, $A_{\text{perp}}{}^n(x+dx)$,

$$z^n{}_{,\beta} A^\beta(x) = A^n(x) = A^n(x+dx) = A_{\tan}{}^\alpha(x+dx) z^n{}_{,\alpha}(x+dx) + A_{\text{perp}}{}^n(x+dx) . \tag{3.36}$$

Since any vector $z^n{}_{,\alpha}$ is contained in the physical hypersurface, this vector is perpendicular to the perpendicular component

$$A_{\text{perp}}{}^n(x+dx) z_{n,\alpha}(x+dx) = 0 .$$

Thus, by multiplying (3.36) with $z_{n,\alpha}(x+dx)$, and using (3.35), we obtain

$$A^{\beta}(x)z^{n}{}_{,\beta}z_{n,\alpha}(x+\mathrm{d}x)=A_{\tan}{}^{\beta}(x+\mathrm{d}x)z^{n}{}_{,\beta}(x+\mathrm{d}x)z_{n,\alpha}(x+\mathrm{d}x)$$
$$=A_{\tan}{}^{\beta}(x+\mathrm{d}x)g_{\alpha\beta}(x+\mathrm{d}x)=A_{\tan\alpha}(x+\mathrm{d}x),$$

as the first term of this expression takes the form

$$A^{\beta}(x)z^{n}{}_{,\beta}(x)z_{n,\alpha}(x+\mathrm{d}x)=A^{\beta}(x)z^{n}{}_{,\beta}(x)\left\lfloor z_{n,\alpha}(x)+z_{n,\alpha\gamma}(x)\mathrm{d}x^{\gamma}\right\rfloor$$
$$=g_{\alpha\beta}(x)A^{\beta}(x)+A^{\beta}(x)z^{n}{}_{,\beta}(x)z_{n,\alpha\gamma}(x)\mathrm{d}x^{\gamma}$$
$$=A_{\alpha}(x)+A^{\beta}(x)\Gamma_{\beta\alpha\gamma}(x)\mathrm{d}x^{\gamma}$$

From (3.36) with these expressions, we find that by a parallel displacement of a vector in the total universe, which does not include any physical effect, we obtain a variation of this vector only due to the curvature of the coordinates,

$$\delta A_{\alpha}(x)=A_{\tan\ \alpha}(x+\mathrm{d}x)-A_{\alpha}(x)=A^{\beta}(x)\Gamma_{\beta\alpha\gamma}(x)\mathrm{d}x^{\gamma}, \qquad \textbf{(3.37)}$$

depending on the Christoffel symbol:

$$\Gamma_{\beta\alpha\gamma}(x)=z^{n}{}_{,\beta}(x)z_{n,\alpha\gamma}(x). \qquad \textbf{(3.38)}$$

This means that, generally, the total variation of a vector, $A_{\mu,\nu}$, includes the variation of some physical effect, $A_{\mu:\nu}$, and the variation (3.37), $\dfrac{\delta A_{\mu}(x)}{\mathrm{d}x^{\nu}}$, due to the parallel displacement in the total space, which does not include any physical effect:

$$A_{\mu,\nu}=A_{\mu:\nu}+A^{\alpha}\Gamma_{\alpha\mu\nu}. \qquad \textbf{(3.39)}$$

It is interesting that the Christoffel symbol (3.38) can be obtained as a function of the metric tensor (3.35), by cyclic permutations of the indices,

$$g_{\alpha\beta,\gamma} = z_{n,\alpha\gamma} z^{n}{}_{,\beta} + \underline{z_{n,\alpha} z^{n}{}_{,\beta\gamma}}$$

$$g_{\beta\gamma,\alpha} = \underline{\underline{z_{n,\beta\alpha} z^{n}{}_{,\gamma}}} + z_{n,\beta} z^{n}{}_{,\gamma\alpha}$$

$$g_{\gamma\alpha,\beta} = \underline{z_{n,\gamma\beta} z^{n}{}_{,\alpha}} + \underline{\underline{z_{n,\gamma} z^{n}{}_{,\alpha\beta}}},$$

up-down interchanges of the dummy indices of the universal coordinates, and by taking into account the symmetry of the metric tensor. By summing the first two equations and subtracting the third, we obtain a symmetric expression with respect to the last two indices:

$$\Gamma_{\beta\alpha\gamma}(x) = z^{n}{}_{,\beta}(x) z_{n,\alpha\gamma}(x) = \frac{1}{2}\left(g_{\beta\alpha,\gamma} + g_{\beta\gamma,\alpha} - g_{\gamma\alpha,\beta}\right) = \Gamma_{\beta\gamma\alpha}(x). \quad \textbf{(3.40)}$$

Due to the symmetry of the metric tensor, from this expression, we obtain the relation

$$\begin{aligned}\Gamma_{\alpha\beta\gamma}(x) + \Gamma_{\beta\alpha\gamma}(x) &= \frac{1}{2}\left(g_{\beta\alpha,\gamma} + \underline{g_{\beta\gamma,\alpha}} - \underline{\underline{g_{\gamma\alpha,\beta}}}\right) + \frac{1}{2}\left(g_{\alpha\beta,\gamma} + \underline{\underline{g_{\alpha\gamma,\beta}}} - \underline{g_{\gamma\beta,\alpha}}\right) \\ &= g_{\alpha\beta,\gamma}.\end{aligned} \quad \textbf{(3.41)}$$

A physical effect is described by a covariant derivative, which according to (3.39), besides the ordinary derivative, includes additional terms proportional to the components of this vector with the Christoffel symbols as coefficients:

$$A_{\mu:\nu} = A_{\mu,\nu} - A^{\alpha}\Gamma_{\alpha\mu\nu}. \quad \textbf{(3.42)}$$

We notice that, by multiplying this expression with the differential of the coordinate vector and (3.37), we obtain a vector,

$$\begin{aligned}A_{\mu:\nu}\mathrm{d}x^{\nu} &= A_{\mu,\nu}(x) - A^{\alpha}(x)\Gamma_{\alpha\mu\nu}(x)\mathrm{d}x^{\nu} \\ &= A_{\mu}(x+\mathrm{d}x) - A_{\mu}(x) - \delta A_{\mu}.\end{aligned} \quad \textbf{(3.43)}$$

According to the quotient theorem demonstrated at the end of the preceding section, the covariant derivative $A_{\mu;\nu}$ is a tensor, unlike the ordinary derivative, which is not a tensor. Really, the relativistic transformation of an ordinary derivative,

$$A_{\mu',\nu'} = \left(A_\rho x^\rho_{,\mu'}\right)_{,\nu'} = A_{\rho,\sigma} x^\sigma_{,\nu'} x^\rho_{,\mu'} + A_\rho x^\rho_{,\mu'\nu'},$$

besides the first term specific to a tensor transformation, includes an additional term. Since the expression (3.42) is a tensor, as its first term is not a tensor, its second term $A^\alpha(x)\Gamma_{\alpha\mu\nu}(x)$ is also a nontensor. Since this term includes a vector as a factor, according to the quotient theorem, the Christoffel symbol is also a nontensor.

From (3.37), by an up-down interchange of the index β, we obtain the variation of a covariant vector as a function of its covariant components,

$$\delta A_\alpha(x) = A_\beta(x)\Gamma^\beta_{\alpha\gamma}(x)\mathrm{d}x^\gamma, \qquad \textbf{(3.44)}$$

and the second kind Christoffel symbol with the symmetry relation according to (3.40):

$$\Gamma^\beta_{\alpha\gamma}(x) = g^{\beta\mu}\Gamma_{\mu\alpha\gamma}(x) = \Gamma^\beta_{\gamma\alpha}(x). \qquad \textbf{(3.45)}$$

From the derivative of (3.15),

$$\left(g^{\alpha\mu} g_{\mu\nu}\right)_{,\sigma} = g^{\alpha\mu}{}_{,\sigma} g_{\mu\nu} + g^{\alpha\mu} g_{\mu\nu,\sigma} = \left(\delta^\alpha_\nu\right)_{,\sigma} = 0,$$

and the multiplication with $g^{\beta\nu}$,

$$g^{\alpha\mu}{}_{,\sigma} g^{\beta\nu} g_{\mu\nu} + g^{\alpha\mu} g^{\beta\nu} g_{\mu\nu,\sigma} = 0$$

$$g^{\alpha\mu}{}_{,\sigma} \delta^\beta_\mu + g^{\alpha\mu} g^{\beta\nu} g_{\mu\nu,\sigma} = 0,$$

we obtain the derivative of the raising metric tensor as a function of the derivative of the lowering tensor:

$$g^{\alpha\beta}{}_{,\sigma} = -g^{\alpha\mu} g^{\beta\nu} g_{\mu\nu,\sigma}. \tag{3.46}$$

By multiplication with an external product of two vectors A and B, we obtain the relation

$$A_\alpha A_\beta g^{\alpha\beta}{}_{,\sigma} = -A^\mu A^\nu g_{\mu\nu,\sigma}. \tag{3.47}$$

With this relation and (3.37), (3.44) and (3.38), we find that by a parallel displacement, the amplitude of a vector remains constant:

$$\begin{aligned}\delta\left(A^\nu A_\nu\right) = \delta\left(g^{\mu\nu} A_\mu A_\nu\right) &= g^{\mu\nu} A_\mu \delta A_\nu + g^{\mu\nu} A_\nu \delta A_\mu + A_\mu A_\nu g^{\mu\nu}{}_{,\sigma} \mathrm{d}x^\sigma \\ &= A^\nu \delta A_\nu + A^\mu \delta A_\mu + A_\alpha A_\beta g^{\alpha\beta}{}_{,\sigma} \mathrm{d}x^\sigma \\ &= A^\nu A^\mu \Gamma_{\mu\nu\sigma} \mathrm{d}x^\sigma + A^\mu A^\nu \Gamma_{\nu\mu\sigma} \mathrm{d}x^\sigma + A_\alpha A_\beta g^{\alpha\beta}{}_{,\sigma} \mathrm{d}x^\sigma \\ &= A^\nu A^\mu g_{\mu\nu,\sigma} \mathrm{d}x^\sigma + A_\alpha A_\beta g^{\alpha\beta}{}_{,\sigma} \mathrm{d}x^\sigma \\ &= A^\nu A^\mu g_{\mu\nu,\sigma} \mathrm{d}x^\sigma - A^\mu A^\nu g_{\mu\nu,\sigma} \mathrm{d}x^\sigma \\ &= 0.\end{aligned} \tag{3.48}$$

From (3.21), we find that the constancy of the length of a vector for a parallel displacement, leads to the constancy of the scalar product for a parallel displacement:

$$\delta\left(A^\nu B_\nu\right) = \delta\left(A_\nu B^\nu\right) = 0. \tag{3.49}$$

From this expression,

$$\begin{aligned}A_\nu \delta B^\nu &= -B^\nu \delta A_\nu = -B^\nu \Gamma^\mu_{\nu\sigma} A_\mu \delta x^\sigma \\ &= -B^\mu \Gamma^\nu_{\mu\sigma} A_\nu \delta x^\sigma\end{aligned}$$

we obtain an expression similar to (3.44), for the variation of a contravariant vector by a parallel displacement:

$$\delta B^\nu = -\Gamma^\nu_{\mu\sigma} B^\mu \delta x^\sigma. \tag{3.50}$$

With the second kind Christoffel symbol (3.45), from (3.42), we obtain the covariant derivative of a covariant vector as a function of its covariant components:

$$A_{\mu:\nu} = A_{\mu,\nu} - \Gamma^{\alpha}_{\mu\nu} A_{\alpha} . \tag{3.51}$$

With this expression, we obtain the covariant derivative of the external product of two vectors

$$\begin{aligned}\left(A_{\mu}B_{\nu}\right)_{:\sigma} &= \left(A_{\mu,\sigma} - \Gamma^{\alpha}_{\mu\sigma}A_{\alpha}\right)B_{\nu} + A_{\mu}\left(B_{\nu,\sigma} - \Gamma^{\alpha}_{\nu\sigma}B_{\alpha}\right) \\ &= \left(A_{\mu}B_{\nu}\right)_{,\sigma} - \Gamma^{\alpha}_{\mu\sigma}A_{\alpha}B_{\nu} - \Gamma^{\alpha}_{\nu\sigma}A_{\mu}B_{\alpha} ,\end{aligned} \tag{3.52}$$

which can be generalized for a tensor

$$T_{\mu\nu:\sigma} = T_{\mu\nu,\sigma} - \Gamma^{\alpha}_{\mu\sigma}T_{\alpha\nu} - \Gamma^{\alpha}_{\nu\sigma}T_{\mu\alpha} ,$$

Evidently, this expression can be generalized for a tensor with an arbitrary number of indices:

$$T_{\mu\nu\rho...:\sigma} = T_{\mu\nu\rho...,\sigma} - \Gamma^{\alpha}_{\mu\sigma}T_{\alpha\nu\rho...} - \Gamma^{\alpha}_{\nu\sigma}T_{\mu\alpha\rho...} - \Gamma^{\alpha}_{\rho\sigma}T_{\mu\nu\alpha...} - ... \tag{3.53}$$

In agreement with (3.49), for a scalar, this expression is

$$S_{:\sigma} = S_{,\sigma} \tag{3.54}$$

From (3.53) with (3.41), we find that the covariant derivative of the metric tensor is null:

$$g_{\mu\nu:\sigma} = g_{\mu\nu,\sigma} - \Gamma^{\alpha}_{\mu\sigma}g_{\alpha\nu} - \Gamma^{\alpha}_{\nu\sigma}g_{\mu\alpha} = g_{\mu\nu,\sigma} - \Gamma_{\nu\mu\sigma} - \Gamma_{\mu\nu\sigma} = 0 . \tag{3.55}$$

To derive an expression similar to (3.51) for a contravariant vector, we consider the covariant derivative of the scalar product of two vectors,

$$\begin{aligned}\left(A^{\mu}B_{\mu}\right)_{:\nu} &= A^{\mu}{}_{:\nu}B_{\mu} + A^{\mu}B_{\mu:\nu} \\ &= A^{\mu}{}_{:\nu}B_{\mu} + A^{\mu}\left(\underline{B_{\mu,\nu}} - \Gamma^{\alpha}_{\mu\nu}B_{\alpha}\right),\end{aligned} \tag{3.56}$$

as the ordinary derivative is

$$\left(A^{\mu}B_{\mu}\right)_{,\nu}=A^{\mu}{}_{,\nu}B_{\mu}+\underline{A^{\mu}B_{\mu,\nu}}\,.$$

From the equality of these derivatives according to (3.54), with an interchange of dummy indices we obtain the equation

$$A^{\mu}{}_{,\nu}B_{\mu}=A^{\mu}{}_{:\nu}B_{\mu}-A^{\mu}\Gamma^{\alpha}_{\mu\nu}B_{\alpha}=A^{\mu}{}_{:\nu}B_{\mu}-A^{\alpha}\Gamma^{\mu}_{\alpha\nu}B_{\mu},$$

which, being valid for any vector B_{μ}, leads to the expression:

$$A^{\mu}{}_{:\nu}=A^{\mu}{}_{,\nu}+\Gamma^{\mu}_{\alpha\nu}A^{\alpha}\,. \tag{3.57}$$

With (3.51) and (3.57), for the scalar product, we obtain an equation of the general form (3.54):

$$\begin{aligned}\left(A^{\mu}B_{\mu}\right)_{:\nu}&=A^{\mu}{}_{:\nu}B_{\mu}+A^{\mu}B_{\mu:\nu}\\&=\left(A^{\mu}{}_{,\nu}+\Gamma^{\mu}_{\alpha\nu}A^{\alpha}\right)B_{\mu}+A^{\mu}\left(B_{\mu,\nu}-\Gamma^{\alpha}_{\mu\nu}B_{\alpha}\right)\\&=A^{\mu}{}_{,\nu}B_{\mu}+A^{\mu}B_{\mu,\nu}\\&=\left(A^{\mu}B_{\mu}\right)_{,\nu}\,.\end{aligned} \tag{3.58}$$

We notice that this theory is based on equation (3.35) of the metric tensor as a sum over the extra-coordinates of the total system. Of course, one could consider the total system reduced to the physical system, with a similar equation to (3.35), but over the physical coordinates, which, in this case, takes the form of a tensor transformation,

$$g_{\alpha\beta}=h_{nm}z^{n}{}_{,\alpha}z^{m}{}_{,\beta}=g_{\alpha'\beta'}x^{\alpha'}{}_{,\alpha}x^{\beta'}{}_{,\beta}\,.$$

This means that the gravitational dynamics could be described in a flat system, with constant metric elements $g_{\alpha'\beta'}=h_{nm}$, used throughout the demonstrations above. Since such a description is impossible, sa curvature of the physical system the inertial-gravitational dynamics is conceivable only in a larger total system, with a larger number of dimensions.

3.3. THE MASS CONSERVATION AS A RELATIVISTIC INVARIANCE

We consider the mass M in the volume of a quantum particle, as an invariant integral of the matter density $\rho\left(x^{\mu}\right)$ for a transformation between two systems of coordinates,

$$\int \rho\left(x^{\mu'}\right) \mathrm{d}x^{0'} \mathrm{d}x^{1'} \mathrm{d}x^{2'} \mathrm{d}x^{3'} = \int \rho\left(x^{\mu}\right) J \, \mathrm{d}x^{0} \mathrm{d}x^{1} \mathrm{d}x^{2} \mathrm{d}x^{3} = M, \tag{3.59}$$

which depends on the Jacobian

$$J = \frac{\partial\left(x^{0'} x^{1'} x^{2'} x^{3'}\right)}{\partial\left(x^{0} x^{1} x^{2} x^{3}\right)} = \mathrm{Det}\left(x^{\mu'}{}_{,\alpha}\right).$$

We notice that the elements of this Jacobian are the elements of the tensor transformation. From the transformation of the metric tensor,

$$g_{\alpha\beta} = x_{,\alpha}{}^{\mu'} g_{\mu'\nu'} x^{\nu'}{}_{,\beta},$$

we obtain the relation between the corresponding determinants

$$g = g' J^2 .$$

Since the determinants of the metric tensor are negative, we obtain the Jacobian

$$J = \frac{\sqrt{-g}}{\sqrt{-g'}} \tag{3.60}$$

Thus, the invariance relation (3.59) takes the form

$$\int \rho\left(x^{\mu'}\right) \sqrt{-g'} \mathrm{d}x^{0'} \mathrm{d}x^{1'} \mathrm{d}x^{2'} \mathrm{d}x^{3'} = \int \rho\left(x^{\mu}\right) \sqrt{-g} \, \mathrm{d}x^{0} \mathrm{d}x^{1} \mathrm{d}x^{2} \mathrm{d}x^{3}, \tag{3.61}$$

as an integral of the quantity $\rho\left(x^{\mu}\right)\sqrt{-g}$, called scalar density. We consider a matter flow four-vector

$$J^{\mu} = \rho v^{\mu}, \tag{3.62}$$

with a null covariant divergence:

$$J^{\mu}{}_{:\mu} = 0. \tag{3.63}$$

With the expression (3.57) of the covariant derivative, this covariant divergence takes the form

$$J^{\mu}{}_{:\mu} = J^{\mu}{}_{,\mu} + \Gamma^{\mu}_{\nu\mu} J^{\nu} = J^{\nu}{}_{,\nu} + \Gamma^{\mu}_{\nu\mu} J^{\nu}. \tag{3.64}$$

With the expression (3.40) of the first kind Christoffel symbol, the derivative of the determinant of the metric matrix

$$g_{,\nu} = g g^{\mu\lambda} g_{\lambda\mu,\nu}, \tag{3.65}$$

and the symmetry of the metric tensor, the second kind Christoffel symbol in (3.64) is

$$\begin{aligned}\Gamma^{\mu}_{\nu\mu} &= \frac{1}{2} g^{\mu\lambda}\left(\underline{g_{\lambda\nu,\mu}} + g_{\lambda\mu,\nu} - \underline{g_{\mu\nu,\lambda}}\right) = \frac{1}{2} g^{\mu\lambda} g_{\lambda\mu,\nu} = \frac{1}{2} g^{-1} g_{,\nu} \\ &= \frac{1}{2}(-g)^{-1}(-g)_{,\nu} = \frac{1}{2}\left[\ln(-g)\right]_{,\nu} = \frac{\left(\sqrt{-g}\right)_{,\nu}}{\sqrt{-g}} = \left(\ln\sqrt{-g}\right)_{,\nu}.\end{aligned} \tag{3.66}$$

With this expression, the covariant divergence (3.64) takes the form

$$J^{\mu}{}_{:\mu}\sqrt{-g} = \left(J^{\nu}\sqrt{-g}\right)_{,\nu}. \tag{3.67}$$

With this expression and equation (3.63) integrated over a spatial volume V , from the Gauss formula, we obtain a conservation equation for the scalar density and flow:

$$\frac{\partial}{c\partial t}\int_{V} \rho v^{0}\sqrt{-g}\mathrm{d}^{3}x = -\oint_{S_V} \rho v^{m}\sqrt{-g}\mathrm{d}^{2}x^{m}. \tag{3.68}$$

In the non-relativistic case, $v^{m} << v^{0} \simeq 1$,

$$J^0 = \rho v^0 \simeq \rho\,,\quad J^m = \rho v^m \doteq \rho\frac{\vec{v}}{c}\,,$$

the conservation equation (3.68) takes the classical form:

$$\frac{\partial}{\partial t}\int_V \rho \mathrm{d}^3\vec{r} = -\oint_{S_V} \rho\vec{v}\mathrm{d}^2\vec{r}\,. \tag{3.69}$$

3.4. HARMONIC OSCILLATIONS AND METRIC TENSOR DENSITY

A harmonic oscillation of a scalar potential $V\left(x^0,x^1,x^2,x^3\right)$ is described by the d'Alembert equation, which in a flat space,

$$g^{\mu\nu} = \begin{pmatrix} 1 & 0 & 0 & 0 \\ 0 & -1 & 0 & 0 \\ 0 & 0 & -1 & 0 \\ 0 & 0 & 0 & -1 \end{pmatrix}, \tag{3.70}$$

is of the form:

$$\Box V = V_{,00} - V_{,11} - V_{,22} - V_{,33} = g^{\mu\nu}V_{,\mu\nu}\,.$$

As we showed in section 3.2, for a physical phenomenon in a curved space, instead of the ordinary derivatives, one has to consider the covariant derivatives:

$$g^{\mu\nu}V_{;\mu\nu} = g^{\mu\nu}\left(V_{,\mu\nu} - \Gamma^{\alpha}_{\mu\nu}V_{,\alpha}\right) = 0\,. \tag{3.71}$$

We notice that this equation takes a simpler form for a coordinate oscillation,

$$x^{\lambda}{}_{,\mu\nu} = \left(x^{\lambda}{}_{,\mu}\right)_{,\nu} = \left(\delta^{\lambda}_{\mu}\right)_{,\nu} = 0\,.$$

In this case, the dynamic equation (3.71) is

$$g^{\mu\nu}\Gamma^{\lambda}_{\mu\nu} = 0\,. \tag{3.72}$$

We can reduce this equation to a more physically understandable form by considering the relation (3.41) between the Christoffel symbol and the metric tensor:

$$g_{\mu\nu,\sigma} = \Gamma_{\mu\nu\sigma} + \Gamma_{\nu\mu\sigma}.$$

With this expression, the derivative (3.46) of a contravariant metric tensor,

$$g^{\mu\nu}{}_{,\sigma} = -g^{\mu\alpha} g^{\nu\beta} g_{\alpha\beta,\sigma},$$

takes the form

$$g^{\mu\nu}{}_{,\sigma} = -g^{\mu\alpha} g^{\nu\beta} \left(\Gamma_{\alpha\beta\sigma} + \Gamma_{\beta\alpha\sigma} \right) = -g^{\nu\beta} \Gamma^{\mu}_{\beta\sigma} - g^{\mu\alpha} \Gamma^{\nu}_{\alpha\sigma}. \tag{3.73}$$

as the derivative of the metric tensor density

$$\left(g^{\mu\nu} \sqrt{-g} \right)_{,\sigma} = g^{\mu\nu}{}_{,\sigma} \sqrt{-g} + g^{\mu\nu} \left(\sqrt{-g} \right)_{,\sigma}. \tag{3.74}$$

With the derivative of the metric density (3.66),

$$\left(\sqrt{-g} \right)_{,\nu} = \sqrt{-g}\, \Gamma^{\mu}_{\nu\mu},$$

and (3.73), from (3.74), we obtain the expression

$$\left(g^{\mu\nu} \sqrt{-g} \right)_{,\sigma} = \left(-g^{\nu\beta} \Gamma^{\mu}_{\beta\sigma} - g^{\mu\alpha} \Gamma^{\nu}_{\alpha\sigma} + g^{\mu\nu} \Gamma^{\beta}_{\sigma\beta} \right) \sqrt{-g}, \tag{3.75}$$

which by a contraction $\sigma = \nu$ reduces to the simpler form:

$$\left(g^{\mu\nu} \sqrt{-g} \right)_{,\nu} = -g^{\nu\beta} \Gamma^{\mu}_{\beta\nu} \sqrt{-g}. \tag{3.76}$$

With this expression, and the symmetry of the metric tensor, or of the Christoffel symbol, the condition (3.72) of a harmonic oscillation takes the explicit form of a null divergence of the metric tensor density:

$$\left(g^{\mu\nu} \sqrt{-g} \right)_{,\nu} = 0. \tag{3.77}$$

3.5. GAUSS AND STOKES THEOREMS WITH COVARIANT DERIVATIVES

As we have previously showed, the external physical fields are described by covariant derivatives. According to (3.57) with (3.66), the covariant divergence of a contravariant vector A^μ is of the form

$$A^\mu{}_{:\mu} = A^\mu{}_{,\mu} + \Gamma^\mu_{\nu\mu} A^\nu = A^\mu{}_{,\mu} + \sqrt{-g}^{-1} \sqrt{-g}_{,\nu} A^\nu . \quad \textbf{(3.78)}$$

Thus, we obtain an expression similar to (3.67) of the covariant divergence as a function of an ordinary divergence of a vector:

$$A^\mu{}_{:\mu} \sqrt{} = \left(A^\mu \sqrt{-g} \right)_{,\mu} . \quad \textbf{(3.79)}$$

By integrating this divergence density over a four-dimensional time-space volume Ω, we obtain a generalization of the Gauss theorem for a four-dimensional space:

$$\int_\Omega A^\mu{}_{:\mu} \sqrt{-g} \mathrm{d}^4 x = \oint_{\Sigma_\Omega} A^\mu \sqrt{-g} \mathrm{d}^3 x_\mu , \quad \textbf{(3.80)}$$

where $\mathrm{d}^3 x_\mu$ is a differential element the hypersurface Σ_Ω in the direction of the considered vector A^μ. Integrating the divergence density (3.79) over a three-dimensional volume of the space, V , with the Gauss theorem, we obtain the equation

$$\int_V A^\mu{}_{:\mu} \sqrt{-g} \mathrm{d}^3 x = \frac{\partial}{c\partial t} \int_V A^0 \sqrt{-g} \mathrm{d}^3 x + \oint_{S_V} A^i \sqrt{-g} \mathrm{d}^2 x_i , \quad \textbf{(3.81)}$$

as a generalization of the conservation equation (3.68) for the creation-annihilation of a fluid with a scalar density A^0 and a flow density A^i.

It is interesting the generalization of the Gauss formula for an antisymmetric tensor $F^{\mu\nu} = -F^{\nu\mu}$. In this case, in the expression of the covariant divergence

$$F^{\mu\nu}{}_{;\nu} = F^{\mu\nu}{}_{,\nu} + \Gamma^{\mu}_{\nu\rho} F^{\rho\nu} + \Gamma^{\nu}_{\nu\rho} F^{\mu\rho},$$

due to the symmetry of the Christoffel symbol, $\Gamma^{\mu}_{\nu\rho} = \Gamma^{\mu}_{\rho\nu}$, and the anti-symmetry of the considered tensor, $F^{\rho\nu} = -F^{\nu\rho}$, the second term vanishes, as the last term, including a Christoffel symbol with two contracted indices, is of the form:

$$\Gamma^{\nu}_{\nu\rho} = \Gamma^{\nu}_{\rho\nu} = \sqrt{-g}^{-1} \sqrt{-g}_{,\rho}.$$

Thus, for an antisymmetric tensor we obtain an equation similar to the vector equation (3.79),

$$F^{\mu\nu}{}_{;\nu} \sqrt{-g} = \left(F^{\mu\nu} \sqrt{-g}\right)_{,\nu}, \tag{3.82}$$

a generalization of the Gauss equation (3.80),

$$\int_{\Omega} F^{\mu\nu}{}_{;\nu} \sqrt{-g} \mathrm{d}^4 x = \oint_{\Sigma_\Omega} F^{\mu\nu} \sqrt{-g} \mathrm{d}^3 x_\nu, \tag{3.83}$$

and a generalization of the fluid creation-annihilation equation (3.81),

$$\int_{V} F^{\mu\nu}{}_{;\nu} \sqrt{-g} \mathrm{d}^3 x = \frac{\partial}{c\partial t} \int_{V} F^{\mu 0} \sqrt{-g} \mathrm{d}^3 x + \oint_{S_V} F^{\mu i} \sqrt{-g} \mathrm{d}^2 x_i. \tag{3.84}$$

For a symmetric tensor $Y^{\mu\nu} = Y^{\nu\mu}$, a similar equation can be obtained, but with additional terms. For this tensor, we consider the covariant derivative of the form

$$Y_{\mu}{}^{\nu}{}_{;\sigma} = Y_{\mu}{}^{\nu}{}_{,\sigma} - \Gamma^{\alpha}_{\mu\sigma} Y_{\alpha}{}^{\nu} + \Gamma^{\nu}_{\alpha\sigma} Y_{\mu}{}^{\alpha},$$

whereby after an index contraction, $\sigma = \nu$, we obtain the last Christoffel symbol as a function only of the metric density $\sqrt{-g}$, $\Gamma^{\nu}_{\alpha\nu} = \sqrt{-g}^{-1} \sqrt{-g}_{,\alpha}$, as in the second term, the considered tensor $Y^{\alpha\nu}$ is obtained with the coefficient $\Gamma_{\alpha\mu\nu}$, expressible only by the metric tensor:

$$Y_{\mu}{}^{\nu}{}_{;\nu} = Y_{\mu}{}^{\nu}{}_{,\nu} - \Gamma^{\alpha}_{\mu\nu} Y_{\alpha}{}^{\nu} + \Gamma^{\nu}_{\alpha\nu} Y_{\mu}{}^{\alpha} = \underline{Y_{\mu}{}^{\nu}{}_{,\nu}} - \Gamma_{\alpha\mu\nu} Y^{\alpha\nu} + \underline{\sqrt{-g}^{-1} \sqrt{-g}_{,\alpha} Y_{\mu}{}^{\alpha}}.$$

By the symmetry of $Y^{\alpha\nu}$, $\Gamma_{\alpha\mu\nu}$ can be replaced by a sum obtainable from equation (3.41),

$$\Gamma_{\alpha\mu\nu}Y^{\alpha\nu}=\frac{1}{2}\left(\Gamma_{\alpha\mu\nu}+\Gamma_{\nu\mu\alpha}\right)Y^{\alpha\nu}=\frac{1}{2}\left(\Gamma_{\alpha\nu\mu}+\Gamma_{\nu\alpha\mu}\right)Y^{\alpha\nu}=\frac{1}{2}g_{\alpha\nu,\mu}Y^{\alpha\nu}.$$

Thus, for a symmetric tensor, we obtain an equation similar to (3.82), but with an additional term:

$$Y_{\mu\ :\nu}^{\ \nu}\sqrt{-g}=\left(Y_{\mu}^{\ \nu}\sqrt{-g}\right)_{,\nu}-\frac{1}{2}g_{\alpha\beta,\mu}Y^{\alpha\beta}\sqrt{-g}\,. \tag{3.85}$$

In this way, the generalized Gauss formula (3.80) for vectors is generalized for symmetric tensors:

$$\int_{\Omega}\left(Y_{\mu\ :\nu}^{\ \nu}+\frac{1}{2}g_{\alpha\beta,\mu}Y^{\alpha\beta}\right)\sqrt{-g}\mathrm{d}^4x=\oint_{\Sigma_\Omega}Y_{\mu}^{\ \nu}\sqrt{-g}\mathrm{d}^3x_{\nu}\,, \tag{3.86}$$

as the creation-annihilation equation for vectors (3.81), is generalized for symmetric tensors:

$$\int_{V}\left(Y_{\mu\ :\nu}^{\ \nu}+\frac{1}{2}g_{\alpha\beta,\mu}Y^{\alpha\beta}\right)\sqrt{-g}\mathrm{d}^3x=\frac{\partial}{c\partial t}\int_{V}Y_{\mu}^{\ 0}\sqrt{-g}\mathrm{d}^3x+\oint_{S_V}Y_{\mu}^{\ i}\sqrt{-g}\mathrm{d}^2x_i\,. \tag{3.87}$$

We notice that the covariant curl of a vector is equal to the ordinary curl:

$$A_{\mu:\nu}-A_{\nu:\mu}=A_{\mu,\nu}-\Gamma^{\alpha}_{\mu\nu}A_{\alpha}-\left(A_{\nu,\mu}-\Gamma^{\alpha}_{\nu\mu}A_{\alpha}\right)=A_{\mu,\nu}-A_{\nu,\mu}\,. \tag{3.88}$$

By integrating this expression on a surface S_Γ with a boundary as a closed curve Γ, we obtain the Stokes theorem:

$$\iint_{S_\Gamma}\left(A_{\mu:\nu}-A_{\nu:\mu}\right)\mathrm{d}x^{\mu}\mathrm{d}x^{\nu}=\iint_{S_\Gamma}\left(A_{\mu,\nu}-A_{\nu,\mu}\right)\mathrm{d}x^{\mu}\mathrm{d}x^{\nu}=\oint_{\Gamma}\left(A_{\mu}\mathrm{d}x^{\mu}+A_{\nu}\mathrm{d}x^{\nu}\right)=2\oint_{\Gamma}A_{\mu}\mathrm{d}x^{\mu}\,, \tag{3.89}$$

where the minus sign comes from the fact that a differential element $-dx^{\nu}$ of the surface S_{Γ}, with the same origin as of dx^{μ}, is opposite to the differential element dx^{ν} of integration on the curve Γ, which is in the continuation dx^{μ}.

3.6. GEODESICS

The geodesic motion is a generalization of the inertial motion of the null acceleration in a flat space, for a particle in a curvilinear system of coordinates describing an inertial-gravitational motion. Such a motion, with coordinates depending on a parameter τ, $z^{\mu}(\tau)$, with a velocity

$$u^{\mu} = \frac{dz^{\mu}}{d\tau}, \tag{3.90}$$

is described by a null covariant acceleration

$$u^{\mu}{}_{;\nu} u^{\nu} = 0, \tag{3.91}$$

which with (3.57) is

$$\left(u^{\mu}{}_{,\nu} + \Gamma^{\mu}_{\nu\sigma} u^{\sigma}\right) u^{\nu} = 0.$$

With (3.90), we obtain the geodesic equation for the acceleration as a derivative of the velocity with a parameter τ,

$$\frac{du^{\mu}}{d\tau} = -\Gamma^{\mu}_{\nu\sigma} u^{\nu} u^{\sigma}, \tag{3.92}$$

or for the acceleration as a second derivative of the coordinates:

$$\frac{d^2 z^{\nu}}{d\tau^2} = -\Gamma^{\mu}_{\nu\sigma} \frac{dz^{\nu}}{d\tau} \frac{dz^{\sigma}}{d\tau}. \tag{3.93}$$

Evidently, for a particle not reaching the light velocity, $ds \neq 0$, these equations take a form depending on the proper time as a parameter:

$$\frac{\mathrm{d}v^{\mu}}{\mathrm{d}s} = -\Gamma^{\mu}_{\nu\sigma} v^{\nu} v^{\sigma}$$
$$\frac{\mathrm{d}^2 z^{\nu}}{\mathrm{d}s^2} = -\Gamma^{\mu}_{\nu\sigma} \frac{\mathrm{d}z^{\nu}}{\mathrm{d}s} \frac{\mathrm{d}z^{\sigma}}{\mathrm{d}s}. \quad \textbf{(3.94)}$$

It is interesting that this equation can be derived from the least action principle, as a minimum value of the action integral between two endpoints P and Q:

$$S = \int_P^Q \mathrm{d}s. \quad \textbf{(3.95)}$$

This means that the variation δS of this action by a variation δz^{λ} of the particle track between the two endpoints is null:

$$\delta S = \int_P^Q \delta \mathrm{d}s = 0. \quad \textbf{(3.96)}$$

By the variation of the squared differential of the time-space interval

$$\mathrm{d}s^2 = g_{\mu\nu} \mathrm{d}z^{\mu} \mathrm{d}z^{\nu},$$

we obtain

$$\begin{aligned} 2\mathrm{d}s\delta\mathrm{d}s &= \mathrm{d}z^{\mu}\mathrm{d}z^{\nu}\delta g_{\mu\nu} + g_{\mu\nu}\mathrm{d}z^{\mu}\delta\mathrm{d}z^{\nu} + g_{\mu\nu}\mathrm{d}z^{\nu}\delta\mathrm{d}z^{\mu} \\ &= \mathrm{d}z^{\mu}\mathrm{d}z^{\nu} g_{\mu\nu,\lambda}\delta z^{\lambda} + 2 g_{\mu\lambda}\mathrm{d}z^{\mu}\delta\mathrm{d}z^{\lambda}. \end{aligned}$$

With the velocity

$$v^{\mu} = \frac{\mathrm{d}z^{\mu}}{\mathrm{d}s},$$

we obtain the time-space interval variation

$$\delta\mathrm{d}s = \left(\frac{1}{2} g_{\mu\nu,\lambda} v^{\mu} v^{\nu} \delta z^{\lambda} + g_{\mu\lambda} v^{\mu} \frac{\mathrm{d}\delta z^{\gamma}}{\mathrm{d}s} \right) \mathrm{d}s.$$

With this expression, the least action equation (3.94) takes the explicit form

$$\delta S = \int_P^Q \left(\frac{1}{2} g_{\mu\nu,\lambda} v^{\mu} v^{\nu} \delta z^{\lambda} + g_{\mu\lambda} v^{\mu} \frac{\mathrm{d}\delta z^{\gamma}}{\mathrm{d}s} \right) \mathrm{d}s = 0.$$

By a partial integration with null coordinate variations for the two end points, $\delta z^{\lambda}(P)=\delta z^{\lambda}(Q)=0$, this equation takes a form

$$\delta S=\int_{P}^{Q}\left[\frac{1}{2}g_{\mu\nu,\lambda}v^{\mu}v^{\nu}-\frac{\mathrm{d}}{\mathrm{d}s}\left(g_{\mu\lambda}v^{\mu}\right)\right]\delta z^{\lambda}\mathrm{d}s=0,$$

which, for arbitrary coordinate variations of the particle track leads to the condition

$$\frac{\mathrm{d}}{\mathrm{d}s}\left(g_{\mu\lambda}v^{\mu}\right)-\frac{1}{2}g_{\mu\nu,\lambda}v^{\mu}v^{\nu}=0.$$

With the expression

$$\frac{\mathrm{d}}{\mathrm{d}s}\left(g_{\mu\lambda}v^{\mu}\right)=g_{\mu\lambda}\frac{\mathrm{d}v^{\mu}}{\mathrm{d}s}+g_{\mu\lambda,\nu}v^{\mu}v^{\nu}=g_{\mu\lambda}\frac{\mathrm{d}v^{\mu}}{\mathrm{d}s}+\frac{1}{2}\left(g_{\lambda\mu,\nu}+g_{\lambda\nu,\mu}\right)v^{\mu}v^{\nu},$$

this equation takes the form

$$g_{\mu\lambda}\frac{\mathrm{d}v^{\mu}}{\mathrm{d}s}=-\frac{1}{2}\left(g_{\lambda\mu,\nu}+g_{\lambda\nu,\mu}-g_{\mu\nu,\lambda}\right)v^{\mu}v^{\nu}.$$

With the expression (3.40) of the first kind Christoffel symbol, we obtain the equation

$$g_{\mu\lambda}\frac{\mathrm{d}v^{\mu}}{\mathrm{d}s}=-\Gamma_{\lambda\mu\nu}v^{\mu}v^{\nu}.$$

By multiplication with the raising metric tensor $g^{\sigma\lambda}$,

$$g^{\sigma\lambda}g_{\mu\lambda}\frac{\mathrm{d}v^{\mu}}{\mathrm{d}s}=\delta_{\mu}^{\sigma}\frac{\mathrm{d}v^{\mu}}{\mathrm{d}s}=-g^{\sigma\lambda}\Gamma_{\lambda\mu\nu}v^{\mu}v^{\nu},$$

we obtain the geodesic equation

$$\frac{\mathrm{d}v^{\sigma}}{\mathrm{d}s}=-\Gamma_{\mu\nu}^{\sigma}v^{\mu}v^{\nu},$$

which is of the form (3.92).

3.7. THE CURVATURE TENSOR

As we have showed above, the covariant differential of a vector A_ν, specific to a physical effect, is obtained from the ordinary differential $\mathrm{d}A_\nu$, by subtracting the variation due to curvature of the coordinate system (3.37)

$$\delta A_\alpha(x) = A_{\tan\,\alpha}(x+\mathrm{d}x) - A_\alpha(x),$$

obtained by a parallel transportation of the vector $A_n = z_{n,\alpha}A^\alpha$ in the total system, where the physical system is a curved four-dimensional hypersurface $z_n(x)$. According to (3.44), this variation is of the form

$$\delta A_\nu(x) = A_\rho(x)\Gamma^\rho_{\nu\sigma}(x)\mathrm{d}x^\sigma, \quad \textbf{(3.97)}$$

depending on the second kind Christoffel symbol which, according to equations (3.45) and (3.40), is a function of the metric tensor:

$$\Gamma^\rho_{\nu\sigma}(x) = g^{\rho\mu}\Gamma_{\mu\nu\sigma} = \frac{1}{2}g^{\rho\mu}\left(g_{\mu\nu,\sigma} + g_{\mu\sigma,\nu} - g_{\nu\sigma,\mu}\right). \quad \textbf{(3.98)}$$

It is remarkable that, although by a parallel displacement in the total system the covariant vector A_ν has a nonzero variation (3.97), according to (3.48) the vector length $A^\nu A_\nu$ remains unchanged. In the length variation $\delta\left(A^\nu A_\nu\right) = A^\nu\delta A_\nu + A_\nu\delta A^\nu$, the variation $A^\nu\delta A_\nu$ is canceled by variation $A_\nu\delta A^\nu$ as an effect of the coordinate dependence on the metric tensor. We obtain a covariant derivative of the form (3.51):

$$A_{\nu:\rho} = A_{\nu,\rho} - \Gamma^\alpha_{\nu\rho}A_\alpha.$$

Evidently, with this expression the second covariant differentiation $A_{\nu:\rho\sigma}$ is not a symmetric operation, $A_{\nu:\rho\sigma} \neq A_{\nu:\sigma\rho}$, unlike the ordinary differentiation, which is a symmetric operation, $A_{\nu,\rho\sigma} = A_{\nu,\sigma\rho}$. Due to the system curvature, a difference between the two second-order derivatives arise: $A_{\nu:\rho\sigma} - A_{\nu:\sigma\rho} \neq 0$. Since, as we showed in section (3.2), the covariant derivative of a vector is a tensor, according to the general formulas (3.53) and (3.51), we obtain the second derivative of the vector A_ν:

$$\begin{aligned}
A_{\nu:\rho\sigma} &= A_{\nu:\rho,\sigma} - \Gamma^{\beta}_{\nu\sigma} A_{\beta:\rho} - \Gamma^{\beta}_{\rho\sigma} A_{\nu:\beta} \\
&= \left(A_{\nu,\rho} - \Gamma^{\beta}_{\nu\rho} A_{\beta}\right)_{,\sigma} - \Gamma^{\beta}_{\nu\sigma}\left(A_{\beta,\rho} - \Gamma^{\alpha}_{\beta\rho} A_{\alpha}\right) - \Gamma^{\beta}_{\rho\sigma}\left(A_{\nu,\beta} - \Gamma^{\alpha}_{\nu\beta} A_{\alpha}\right) \\
&= A_{\nu,\rho\sigma} - \Gamma^{\beta}_{\nu\rho} A_{\beta,\sigma} - \Gamma^{\beta}_{\nu\sigma} A_{\beta,\rho} - \Gamma^{\beta}_{\rho\sigma} A_{\nu,\beta} \\
&\quad - A_{\alpha}\left(\Gamma^{\alpha}_{\nu\rho,\sigma} - \Gamma^{\beta}_{\nu\sigma}\Gamma^{\alpha}_{\beta\rho} - \underline{\Gamma^{\beta}_{\rho\sigma}\Gamma^{\alpha}_{\nu\beta}}\right).
\end{aligned}$$

By reversing the order of the two derivations,

$$\begin{aligned}
A_{\nu:\sigma\rho} &= A_{\nu,\sigma\rho} - \Gamma^{\beta}_{\nu\sigma} A_{\beta,\rho} - \Gamma^{\beta}_{\nu\rho} A_{\beta,\sigma} - \Gamma^{\beta}_{\sigma\rho} A_{\nu,\beta} \\
&\quad - A_{\alpha}\left(\Gamma^{\alpha}_{\nu\sigma,\rho} - \Gamma^{\beta}_{\nu\rho}\Gamma^{\alpha}_{\beta\sigma} - \underline{\Gamma^{\beta}_{\sigma\rho}\Gamma^{\alpha}_{\nu\beta}}\right),
\end{aligned}$$

we obtain the difference between the two second-order derivatives of the form

$$A_{\nu:\rho\sigma} - A_{\nu:\sigma\rho} = R^{\alpha}_{\nu\rho\sigma} A_{\alpha}, \tag{3.99}$$

depending on the quantity

$$\begin{aligned}
R^{\alpha}_{\nu\rho\sigma} &= -\Gamma^{\alpha}_{\nu\rho,\sigma} - \Gamma^{\beta}_{\nu\rho}\Gamma^{\alpha}_{\beta\sigma} + \Gamma^{\alpha}_{\nu\sigma,\rho} + \Gamma^{\beta}_{\nu\sigma}\Gamma^{\alpha}_{\beta\rho} \\
&= -\langle\rho\sigma\rangle + \Gamma^{\alpha}_{\nu\sigma,\rho} + \Gamma^{\beta}_{\nu\sigma}\Gamma^{\alpha}_{\beta\rho} = -\langle\rho\sigma\rangle + \langle\sigma\rho\rangle \\
&\simeq -\Gamma^{\alpha}_{\nu\rho,\sigma} + \Gamma^{\alpha}_{\nu\sigma,\rho}.
\end{aligned} \tag{3.100}$$

According to the quotient theorem, $R^{\alpha}_{\nu\rho\sigma}$ is a tensor called the curvature tensor, as a difference of two quantities differing only by an interchange of indices, $\langle\rho\sigma\rangle$ and $\langle\sigma\rho\rangle$. The approximate equality corresponds to the neglect of the second-order terms. From this expression, or from (3.99), we notice the curvature tensor is antisymmetric with respect to the last two indices ρ and σ:

$$R^{\alpha}_{\nu\rho\sigma} = -R^{\alpha}_{\nu\sigma\rho}\,. \tag{3.101}$$

By circular permutations of the indices ν, ρ, σ, with the symmetry property (3.45) of the Christoffel symbol,

$$\begin{aligned}
R^{\alpha}_{\nu\rho\sigma} &= \underline{-\Gamma^{\alpha}_{\nu\rho,\sigma}} - \Gamma^{\beta}_{\nu\rho}\Gamma^{\alpha}_{\beta\sigma} + \underline{\underline{\Gamma^{\alpha}_{\nu\sigma,\rho}}} + \underline{\Gamma^{\beta}_{\nu\sigma}\Gamma^{\alpha}_{\beta\rho}}\\
R^{\alpha}_{\rho\sigma\nu} &= -\Gamma^{\alpha}_{\rho\sigma,\nu} - \Gamma^{\beta}_{\rho\sigma}\Gamma^{\alpha}_{\beta\nu} + \underline{\Gamma^{\alpha}_{\rho\nu,\sigma} + \Gamma^{\beta}_{\rho\nu}\Gamma^{\alpha}_{\beta\sigma}}\\
R^{\alpha}_{\sigma\nu\rho} &= \underline{\underline{-\Gamma^{\alpha}_{\sigma\nu,\rho} - \Gamma^{\beta}_{\sigma\nu}\Gamma^{\alpha}_{\beta\rho}}} + \Gamma^{\alpha}_{\sigma\rho,\nu} + \Gamma^{\beta}_{\sigma\rho}\Gamma^{\alpha}_{\beta\nu}\,,
\end{aligned}$$

we obtain the symmetry relation:

$$R^{\alpha}_{\nu\rho\sigma} + R^{\alpha}_{\rho\sigma\nu} + R^{\alpha}_{\sigma\nu\rho} = 0\,. \tag{3.102}$$

We define the covariant curvature tensor

$$R_{\mu\nu\rho\sigma} = g_{\mu\alpha}R^{\alpha}_{\nu\rho\sigma}\,. \tag{3.103}$$

With (3.98) and (3.41), from (3.100) we obtain the explicit expression

$$\begin{aligned}
R_{\mu\nu\rho\sigma} &= g_{\mu\alpha}\left(-\Gamma^{\alpha}_{\nu\rho,\sigma} - \Gamma^{\beta}_{\nu\rho}\Gamma^{\alpha}_{\beta\sigma} + \Gamma^{\alpha}_{\nu\sigma,\rho} + \Gamma^{\beta}_{\nu\sigma}\Gamma^{\alpha}_{\beta\rho}\right)\\
&= -\Gamma_{\mu\nu\rho,\sigma} + g_{\mu\alpha,\sigma}\Gamma^{\alpha}_{\nu\rho} - \Gamma^{\beta}_{\nu\rho}\Gamma_{\mu\beta\sigma} + \Gamma_{\mu\nu\sigma,\rho} - g_{\mu\alpha,\rho}\Gamma^{\alpha}_{\nu\sigma} + \Gamma^{\beta}_{\nu\sigma}\Gamma_{\mu\beta\rho}\\
&= -\Gamma_{\mu\nu\rho,\sigma} + \Gamma_{\mu\nu\sigma,\rho} + \left(\underline{\Gamma_{\mu\alpha\sigma}} + \Gamma_{\alpha\mu\sigma}\right)\Gamma^{\alpha}_{\nu\rho} - \underline{\Gamma^{\beta}_{\nu\rho}\Gamma_{\mu\beta\sigma}} - \left(\underline{\underline{\Gamma_{\mu\alpha\rho}}} + \Gamma_{\alpha\mu\rho}\right)\Gamma^{\alpha}_{\nu\sigma} + \underline{\underline{\Gamma^{\beta}_{\nu\sigma}\Gamma_{\mu\beta\rho}}}\\
&= -\Gamma_{\mu\nu\rho,\sigma} + \Gamma^{\alpha}_{\nu\rho}\Gamma_{\alpha\mu\sigma} + \Gamma_{\mu\nu\sigma,\rho} - \Gamma^{\alpha}_{\nu\sigma}\Gamma_{\alpha\mu\rho}
\end{aligned}$$

which is

$$\begin{aligned} R_{\mu\nu\rho\sigma} &= -\Gamma_{\mu\nu\rho,\sigma} + \Gamma_{\alpha\nu\rho}\Gamma^{\alpha}_{\mu\sigma} + \Gamma_{\mu\nu\sigma,\rho} - \Gamma_{\alpha\nu\sigma}\Gamma^{\alpha}_{\mu\rho} \\ &= \langle\rho\sigma\rangle + \Gamma_{\mu\nu\sigma,\rho} - \Gamma_{\alpha\nu\sigma}\Gamma^{\alpha}_{\mu\rho} = \langle\rho\sigma\rangle - \langle\sigma\rho\rangle \\ &\simeq -\Gamma_{\mu\nu\rho,\sigma} + \Gamma_{\mu\nu\sigma,\rho}, \end{aligned} \tag{3.104}$$

as a difference of two quantities differing only by an interchange of indices, $\langle\rho\sigma\rangle$ and $\langle\rho\sigma\rangle$. From this expression, we notice that the covariant curvature tensor $R_{\mu\nu\rho\sigma}$ satisfies an anti-symmetry relation for these indices, similar to the anti-symmetry relation (3.101) for the contravariant curvature tensor $R^{\alpha}_{\nu\rho\sigma}$:

$$R_{\mu\nu\rho\sigma} = -R_{\mu\nu\sigma\rho}. \tag{3.105}$$

At the same time, with the expression (3.40) of the first kind Christoffel symbol as a function of the metric tensor, for the first-order terms of the covariant curvature tensor (3,104), we obtain an expression antisymmetric with the first two indices μ and ν,

$$\begin{aligned} -\Gamma_{\mu\nu\rho,\sigma} + \Gamma_{\mu\nu\sigma,\rho} &= -\frac{1}{2}\left(g_{\mu\nu,\rho} + g_{\mu\rho,\nu} - g_{\nu\rho,\mu}\right)_{,\sigma} + \frac{1}{2}\left(g_{\mu\nu,\sigma} + g_{\mu\sigma,\nu} - g_{\nu\sigma,\mu}\right)_{,\rho} \\ &= \frac{1}{2}\left(\underline{-g_{\mu\nu,\rho\sigma}} - g_{\mu\rho,\nu\sigma} + g_{\nu\rho,\mu\sigma} + \underline{g_{\mu\nu,\sigma\rho}} + g_{\mu\sigma,\nu\rho} - g_{\nu\sigma,\mu\rho}\right) \\ &= \frac{1}{2}\left(-g_{\mu\rho,\nu\sigma} + g_{\nu\rho,\mu\sigma} - g_{\nu\sigma,\mu\rho} + g_{\mu\sigma,\nu\rho}\right). \end{aligned} \tag{3.106}$$

With an up-down index change, we obtain that also the second-order terms of the curvature tensor (3.104) are antisymmetric with the indices μ and ν:

$$\Gamma_{\alpha\nu\rho}\Gamma^{\alpha}_{\mu\sigma} - \Gamma_{\alpha\nu\sigma}\Gamma^{\alpha}_{\mu\rho} = \Gamma_{\alpha\nu\rho}\Gamma^{\alpha}_{\mu\sigma} - \Gamma^{\alpha}_{\nu\sigma}\Gamma_{\alpha\mu\rho}. \tag{3.107}$$

Thus, the covariant curvature tensor (3.103) is antisymmetric also with the first two indices:

$$R_{\mu\nu\rho\sigma} = -R_{\nu\mu\rho\sigma}. \tag{3.108}$$

We notice that the first-order terms (3.106) of the curvature tensor (3.104) are symmetrical with the index interchange $\mu \leftrightarrow \rho$, $\nu \leftrightarrow \sigma$,

$$\begin{aligned} -\Gamma_{\mu\nu\rho,\sigma} + \Gamma_{\mu\nu\sigma,\rho} &= \frac{1}{2}\left(-\underline{g_{\mu\rho,\nu\sigma}} - \underline{\underline{g_{\nu\sigma,\mu\rho}}} + \underline{g_{\mu\sigma,\nu\rho}} + \underline{\underline{g_{\nu\rho,\mu\sigma}}}\right) \\ &= \frac{1}{2}\left(-\underline{g_{\rho\mu,\sigma\nu}} - \underline{\underline{g_{\sigma\nu,\rho\mu}}} + \underline{\underline{g_{\rho\nu,\sigma\mu}}} + \underline{g_{\sigma\mu,\rho\nu}}\right)' \end{aligned}$$

and that, due to the symmetry of the Christoffel symbols, the second order terms (3.107) of this tensor are also symmetrical for this interchange of indices:

$$\begin{aligned} \Gamma_{\alpha\nu\rho}\Gamma^{\alpha}_{\mu\sigma} - \Gamma_{\alpha\nu\sigma}\Gamma^{\alpha}_{\mu\rho} &= \Gamma_{\alpha\sigma\mu}\Gamma^{\alpha}_{\rho\nu} - \Gamma_{\alpha\sigma\nu}\Gamma^{\alpha}_{\rho\mu} = \Gamma^{\alpha}_{\sigma\mu}\Gamma_{\alpha\rho\nu} - \Gamma_{\alpha\sigma\nu}\Gamma^{\alpha}_{\rho\mu} \\ &= \Gamma_{\alpha\rho\nu}\Gamma^{\alpha}_{\sigma\mu} - \Gamma_{\alpha\sigma\nu}\Gamma^{\alpha}_{\rho\mu}. \end{aligned}$$

We obtain the symmetry relation

$$R_{\mu\nu\rho\sigma} = R_{\rho\sigma\mu\nu}. \tag{3.109}$$

Finally, by multiplying (3.102) with $g_{\mu\alpha}$, we obtain the symmetry relation

$$R_{\mu\nu\rho\sigma} + R_{\mu\rho\sigma\nu} + R_{\mu\sigma\nu\rho} = 0. \tag{3.110}$$

By a detailed analysis of the four symmetry relations (3.105), (3.108), (3.109), and (3,110), one can show that the number of 4^4=256 elements of the curvature tensor reduces to only 20 [9]. For this, first of all, we notice that the 4 elements with all the 4 indices equal are null. A non-zero element has at least 2 different indices, which means that a non-zero curvature elements is obtained only with: (1) 2 different indices, (2) 3 different indices, and (3) all the 4 indices different. It is interesting that by the symmetry relations (3.105), (3.108), and (3.109), any index can be brought in the first position:

$$
\begin{aligned}
&R_{\alpha\beta\gamma\delta} \\
&R_{\alpha\beta\gamma\delta} = -R_{\beta\alpha\gamma\delta} \\
&R_{\alpha\beta\gamma\delta} = R_{\gamma\delta\alpha\beta} \\
&R_{\alpha\beta\gamma\delta} = -R_{\alpha\beta\delta\gamma} = -R_{\delta\gamma\alpha\beta}\,,
\end{aligned}
$$

which means that any index can be brought in any position. For the case (1) of two-different indices, we have $\mathrm{C}_4^2 = \dfrac{4\cdot 3}{1\cdot 2} = 6$ matrix elements. For the case (2) of three different indices, we have $\mathrm{C}_4^3 = \mathrm{C}_4^1 = 4$ matrix elements multiplied with 3, for the three indices doubled to remain to the three different indices,

$$
\begin{array}{lll}
\underline{\alpha}\underline{\alpha}\beta\gamma & \underline{\beta}\alpha\underline{\beta}\gamma & \underline{\gamma}\alpha\beta\underline{\gamma} \\
\underline{\alpha}\underline{\alpha}\gamma\delta & \underline{\gamma}\alpha\underline{\gamma}\delta & \underline{\delta}\alpha\gamma\underline{\delta} \\
\underline{\alpha}\underline{\alpha}\delta\beta & \underline{\delta}\alpha\underline{\delta}\beta & \underline{\beta}\alpha\delta\underline{\beta} \\
\underline{\beta}\underline{\beta}\gamma\delta & \underline{\gamma}\beta\underline{\gamma}\delta & \underline{\delta}\beta\gamma\underline{\delta}\,,
\end{array}
$$

which, can also be understood as the 4 double indices multiplied with the combinations of the other 3, $4\times\mathrm{C}_3^2 = 4\times 3 = 12$:

$$
\begin{array}{lll}
\alpha\alpha\beta\gamma & \alpha\alpha\gamma\delta & \alpha\alpha\delta\beta \\
\beta\beta\alpha\delta & \beta\beta\gamma\alpha & \beta\beta\delta\gamma \\
\gamma\gamma\alpha\beta & \gamma\gamma\beta\delta & \gamma\gamma\delta\alpha \\
\delta\delta\alpha\beta & \delta\delta\beta\gamma & \delta\delta\gamma\alpha\,.
\end{array}
$$

For the case (3) of all the four different indices, we obtain 4!=24 independent matrix elements, which with (3.105), (3.108), and (3.109) are reduced to $24/2^3=3$, and with (3.110) are reduced to only 2 independent matrix elements. Totally, we obtain 6+12+2=20 independent matrix elements.

In a flat space, where the metric elements are constant, which means that the Christoffel symbols are null, all the curvature elements are null. Conversely, we can prove that if the curvature tensor is null, the space is flat. For this, we notice that according to its definition (3.99), if the curvature tensor is null in in the space of a system S, this space is flat. In the following, we show that if a space is flat in a system of reference S, it is flat in any other system of reference S'. For this, we consider the relation between the metric elements of the two systems of reference,

$$g_{\mu\lambda} = g_{\alpha'\beta'} x^{\alpha'}_{,\mu} x^{\beta'}_{,\lambda} \tag{3.111}$$

and the relation between their derivatives:

$$g_{\mu\lambda,\nu} - g_{\alpha'\beta',\nu} x^{\alpha'}_{,\mu} x^{\beta'}_{,\lambda} = g_{\alpha'\beta'} \left(x^{\alpha'}_{,\mu\nu} x^{\beta'}_{,\lambda} + x^{\alpha'}_{,\mu} x^{\beta'}_{,\lambda\nu} \right). \tag{3.112}$$

To calculate the second-order derivatives in this expression, we notice that the first derivative of a scalar is a vector:

$$S_{,\mu} = S_{,\mu'} x^{\mu'}{}_{,\mu}, \tag{3.113}$$

Since in a flat system the covariant derivative of a vector is null, the ordinary derivative in such a system is of the form:

$$A_{\mu,\nu} = A_{\mu:\nu} + \Gamma^{\sigma}_{\mu\nu} A_{\sigma} = \Gamma^{\sigma}_{\mu\nu} A_{\sigma}, \tag{3.114}$$

which means that the second derivative of a scalar in a flat system is of the form

$$S_{,\mu\nu} = \Gamma^{\sigma}_{\mu\nu} S_{,\sigma}. \tag{3.115}$$

With this expression, we obtain the second derivatives of the coordinates in (3.112):

$$x^{\alpha'}_{,\mu\nu} = \Gamma^{\sigma}_{\mu\nu} x^{\alpha'}_{,\sigma}. \tag{3.116}$$

With this expression and equation (3.41), we find that in a flat system the relation (110) takes the form

$$\underline{g_{\mu\lambda,\nu}} - g_{\alpha'\beta',\nu} x^{\alpha'}_{,\mu} x^{\beta'}_{,\lambda} = g_{\alpha'\beta'} \left(\Gamma^{\sigma}_{\mu\nu} x^{\alpha'}_{,\sigma} x^{\beta'}_{,\lambda} + x^{\alpha'}_{,\mu} \Gamma^{\sigma}_{\lambda\nu} x^{\beta'}_{,\sigma} \right)$$
$$= g_{\sigma\lambda} \Gamma^{\sigma}_{\mu\nu} + g_{\mu\sigma} \Gamma^{\sigma}_{\lambda\nu}$$
$$= \Gamma_{\lambda\mu\nu} + \Gamma_{\mu\lambda\nu} = \underline{g_{\mu\lambda,\nu}}\,.$$

Thus, we obtained that if the space is flat in a system of reference S, it is flat in any other system of reference S':

$$g_{\alpha'\beta',\nu} = 0\,.$$

3.8. BIANCI RELATIONS AND EINSTEIN'S LAW OF GRAVITATION

Einstein's remarkable idea was that gravitation means a curvature of the physical space, which can be conceived only in a larger total space. This means that the law of gravitation should be formulated as a differential equation of the curvature tensor $R^{\mu}_{\nu\rho\sigma}$, with a null covariant derivative. To obtain the covariant derivative of a tensor as a function of the curvature tensor, we consider a double covariant derivative of an external product of vectors, and the corresponding expression with the reverseed order of derivation:

$$\left(A_\mu B_\tau\right)_{:\rho\sigma} = \left(A_{\mu:\rho} B_\tau + A_\mu B_{\tau:\rho}\right)_{:\sigma} = A_{\mu:\rho\sigma} B_\tau + \underline{A_{\mu:\rho} B_{\tau:\sigma}} + \underline{\underline{A_{\mu:\sigma} B_{\tau:\rho}}} + A_\mu B_{\tau:\rho\sigma}$$

$$\left(A_\mu B_\tau\right)_{:\sigma\rho} = \left(A_{\mu:\sigma} B_\tau + A_\mu B_{\tau:\sigma}\right)_{:\rho} = A_{\mu:\sigma\rho} B_\tau + \underline{\underline{A_{\mu:\sigma} B_{\tau:\rho}}} + \underline{A_{\mu:\rho} B_{\tau:\sigma}} + A_\mu B_{\tau:\sigma\rho}\,.$$

By subtraction, we obtain the expression

$$\left(A_\mu B_\tau\right)_{:\rho\sigma} - \left(A_\mu B_\tau\right)_{:\sigma\rho} = \left(A_{\mu:\rho\sigma} - A_{\mu:\sigma\rho}\right) B_\tau + A_\mu \left(B_{\tau:\rho\sigma} - B_{\tau:\sigma\rho}\right),$$

which, with (3.99) is

$$\left(A_\mu B_\tau\right)_{:\rho\sigma} - \left(A_\mu B_\tau\right)_{:\sigma\rho} = R^{\alpha}_{\mu\rho\sigma} A_\alpha B_\tau + R^{\alpha}_{\tau\rho\sigma} A_\mu B_\alpha\,.$$

This means that for a tensor, which can always be conceived as a sum of external products of vectors, we obtain the expression

$$T_{\mu\tau:\rho\sigma} - T_{\mu\tau:\sigma\rho} = R^{\alpha}_{\mu\rho\sigma} T_{\alpha\tau} + R^{\alpha}_{\tau\rho\sigma} T_{\mu\alpha} . \quad \textbf{(3.117)}$$

For the tensor of a the covariant derivative of a vector, $T_{\mu\tau} = A_{\mu:\tau}$, and cyclic permutations of the indices τ, ρ, σ, we obtain:

$$A_{\mu:\tau\rho\sigma} - \underline{\underline{A_{\mu:\tau\sigma\rho}}} = R^{\alpha}_{\mu\rho\sigma} A_{\alpha:\tau} + R^{\alpha}_{\tau\rho\sigma} A_{\mu:\alpha}$$

$$\underline{A_{\mu:\rho\sigma\tau}} - A_{\mu:\rho\tau\sigma} = R^{\alpha}_{\mu\sigma\tau} A_{\alpha:\rho} + R^{\alpha}_{\rho\sigma\tau} A_{\mu:\alpha}$$

$$\underline{\underline{A_{\mu:\sigma\tau\rho}}} - \underline{A_{\mu:\sigma\rho\tau}} = R^{\alpha}_{\mu\tau\rho} A_{\alpha:\sigma} + R^{\alpha}_{\sigma\tau\rho} A_{\mu:\alpha} .$$

By the summation of these relations, for the left-hand side we obtain,

$$A_{\mu:\tau\rho\sigma} - A_{\mu:\rho\tau\sigma} + \underline{A_{\mu:\rho\sigma\tau} - A_{\mu:\sigma\rho\tau}} + \underline{\underline{A_{\mu:\sigma\tau\rho} - A_{\mu:\tau\sigma\rho}}} = \left(R^{\alpha}_{\mu\tau\rho} A_{\alpha}\right)_{:\sigma} + \left(R^{\alpha}_{\mu\rho\sigma} A_{\alpha}\right)_{:\tau} + \left(R^{\alpha}_{\mu\sigma\tau} A_{\alpha}\right)_{:\rho}$$

$$= \left(R^{\alpha}_{\mu\tau\rho:\sigma} + R^{\alpha}_{\mu\rho\sigma:\tau} + R^{\alpha}_{\mu\sigma\tau:\rho}\right) A_{\alpha} + R^{\alpha}_{\mu\tau\rho} A_{\alpha:\sigma} + R^{\alpha}_{\mu\rho\sigma} A_{\alpha:\tau} + R^{\alpha}_{\mu\sigma\tau} A_{\alpha:\rho} ,$$

as the right-hand side is

$$R^{\alpha}_{\mu\tau\rho} A_{\alpha:\sigma} + R^{\alpha}_{\mu\rho\sigma} A_{\alpha:\tau} + R^{\alpha}_{\mu\sigma\tau} A_{\alpha:\rho} + \left(R^{\alpha}_{\tau\rho\sigma} + R^{\alpha}_{\rho\sigma\tau} + R^{\alpha}_{\sigma\tau\rho}\right) A_{\mu:\alpha} .$$

Since the last term of this expression is null according to (3.102), as the other terms reduce with the corresponding terms from the left-hand side, we are left with Bianci's relation:

$$R^{\alpha}_{\mu\tau\rho:\sigma} + R^{\alpha}_{\mu\rho\sigma:\tau} + R^{\alpha}_{\mu\sigma\tau:\rho} = 0 , \quad \textbf{(3.118)}$$

as a sum of three covariant derivatives. To obtain a relation with a single covariant derivative, we contract $\tau = \alpha$,

$$R^{\alpha}_{\mu\alpha\rho:\sigma} + R^{\alpha}_{\mu\rho\sigma:\alpha} + R^{\alpha}_{\mu\sigma\alpha:\rho} = 0 ,$$

and multiply with the metric tensor $g^{\mu\rho}$. By taking into account that the covariant derivative of the metric tensor is null, we obtain

$$\left(g^{\mu\rho}R^{\alpha}_{\mu\alpha\rho}\right)_{:\sigma}+\left(g^{\mu\rho}R^{\alpha}_{\mu\rho\sigma}\right)_{:\alpha}+\left(g^{\mu\rho}R^{\alpha}_{\mu\sigma\alpha}\right)_{:\rho}=0. \tag{3.119}$$

In this expression we distinguish a curvature tensor with a contraction of the first and the last indices, called the Ricci's tensor:

$$R_{\nu\rho}=R^{\mu}_{\nu\rho\mu}=g^{\sigma\mu}R_{\mu\nu\rho\sigma}. \tag{3.120}$$

From the symmetry relations (3.103), (3.106), and (3.107),

$$R_{\mu\nu\rho\sigma}=R_{\rho\sigma\mu\nu}=-R_{\rho\sigma\nu\mu}=R_{\sigma\rho\nu\mu},$$

by multiplication with $g^{\sigma\mu}$,

$$g^{\sigma\mu}R_{\mu\nu\rho\sigma}=g^{\sigma\mu}R_{\sigma\rho\nu\mu}=g^{\mu\sigma}R_{\sigma\rho\nu\mu},$$

from the second expression (3.120) we obtain that the Ricci tensor is symmetric:

$$R_{\nu\rho}=R_{\rho\nu}.$$

In the first term of (3.119) we distinguish the scalar

$$R=-g^{\mu\rho}R^{\alpha}_{\mu\alpha\rho}=g^{\mu\rho}R^{\alpha}_{\mu\rho\alpha}=g^{\mu\rho}R_{\mu\rho}, \tag{3.121}$$

which we call the scalar curvature, or the total curvature. The second term of the expression (3.119), is of the form

$$g^{\mu\rho}R^{\alpha}_{\mu\rho\sigma}=g^{\mu\rho}g^{\alpha\beta}R_{\beta\mu\rho\sigma}=g^{\mu\rho}g^{\alpha\beta}R_{\mu\beta\sigma\rho}$$
$$=g^{\alpha\beta}R^{\rho}_{\beta\sigma\rho}=g^{\alpha\beta}R_{\beta\sigma}=R^{\alpha}{}_{\sigma},$$

as the third term is

$$g^{\mu\rho}R^{\alpha}_{\mu\sigma\alpha}=g^{\mu\rho}R_{\mu\sigma}=R^{\rho}{}_{\sigma}.$$

With these expressions, from (3.119) we obtain the equation

$$-R_{:\sigma}+R^{\alpha}{}_{\sigma:\alpha}+R^{\rho}{}_{\sigma:\rho}=2R^{\alpha}{}_{\sigma:\alpha}-R_{:\sigma}=0.$$

By multiplying with the raising tensor $g^{\sigma\beta}$,

$$2R^{\alpha}{}_{\sigma:\alpha}g^{\sigma\beta} - g^{\sigma\beta}R_{:\sigma} = 2R^{\alpha\beta}{}_{:\alpha} - g^{\alpha\beta}R_{:\alpha} = 0,$$

and taking into account the symmetries of the metric and Ricci tensors we obtain the Bianci relation for the Ricci tensor,

$$\left(R^{\sigma\alpha} - \frac{1}{2}g^{\sigma\alpha}R\right)_{:\alpha} = 0, \qquad \textbf{(3.122)}$$

which leads to Einstein's equation of gravitation in vacuum:

$$R^{\sigma\alpha} - \frac{1}{2}g^{\sigma\alpha}R = 0. \qquad \textbf{(3.123)}$$

From the expression (3.100) of the curvature tensor, we obtain an explicit expression for the Ricci tensor:

$$R_{\nu\rho} = R^{\alpha}_{\nu\rho\alpha} = -\Gamma^{\alpha}_{\nu\rho,\alpha} - \Gamma^{\beta}_{\nu\rho}\Gamma^{\alpha}_{\beta\alpha} + \Gamma^{\alpha}_{\nu\alpha,\rho} + \Gamma^{\beta}_{\nu\alpha}\Gamma^{\alpha}_{\beta\rho}. \qquad \textbf{(3.124)}$$

We notice that the first, the second, and the fourth terms are evidently symmetric with the two indices ν and ρ. The symmetry with these indices of the third term comes from the expression (3.66) of the Christoffel symbol with two contracted indices, which leads to the symmetric expression:

$$\Gamma^{\alpha}_{\nu\alpha,\rho} = \left(\ln\sqrt{-g}\right)_{,\nu\rho}.$$

3.9. NEWTON'S GRAVITATION EQUATION AND RED SHIFT

Newton's gravitation equation refers to the gravitational acceleration/intensity in a constant gravitational field,

$$g_{\mu\nu,0} = 0, \; g_{0i} = 0, \; i = 1.2,3, \qquad \textbf{(3.125)}$$

in vacuum, as a function of the distance r from the gravitational center. This acceleration/field intensity is obtainable as an approximation of Einstein's gravitational law (3.123) with its simplest form

$$R_{\mu\nu} = 0. \tag{3.126}$$

For a weak gravitational field, the Ricci tensor

$$R_{\mu\nu} = R^{\alpha}_{\mu\nu\alpha} = g^{\alpha\beta} R_{\beta\mu\nu\alpha}$$

can be considered in the first-order approximation of the expression (3.104)

$$R_{\beta\mu\nu\alpha} \simeq -\Gamma_{\beta\mu\nu,\alpha} + \Gamma_{\beta\mu\alpha,\nu}$$

which, with the expression (3.106) is

$$-\Gamma_{\beta\mu\nu,\alpha} + \Gamma_{\beta\mu\alpha,\nu} = \frac{1}{2}\left(-g_{\beta\nu,\mu\alpha} + g_{\mu\nu,\beta\alpha} - g_{\mu\alpha,\beta\nu} + g_{\beta\alpha,\mu\nu}\right).$$

With these expressions, Einstein's equation (3.126) takes the form

$$g^{\alpha\beta}\left(-g_{\beta\nu,\mu\alpha} + g_{\mu\nu,\beta\alpha} - g_{\mu\alpha,\beta\nu} + g_{\beta\alpha,\mu\nu}\right) = 0,$$

which, for $\mu = \nu = 0$, with the conditions (3.125), is a Laplace equation for the time metric element:

$$g^{ij} g_{00,ji} = 0. \tag{3.127}$$

We consider the time metric element g_{00}, as gravitational potential for a weak gravitational field, in a system of reference where the metric tensor takes a diagonal form depending on the normalized gravitational potential V :

$$g_{00} = \frac{1}{g^{00}} = 1 + 2V \simeq 1, \; g_{11} \simeq -1, \; g_{22} \simeq -1, \; g_{33} = -1. \tag{3.128}$$

Equation (3.127) takes the form of the ordinary Laplace equation for this potential:

$$\nabla^2 V = 0.$$

In a system with spherical symmetry, this equation has a solution depending on the distance r from the gravitational center and an integration constant m,

$$V = -\frac{m}{r}. \tag{3.129}$$

In a gravitational potential, a quantum particle is accelerated according to the geodesic equation (3.94). With the explicit expression (3.40) of the Christoffel symbol, equation (3.94) for the ordinary acceleration of a quantum particle in the proper time takes the form

$$\frac{\mathrm{d}v^i}{\mathrm{d}s} = -\Gamma^i_{00} v^{0^2} = -g^{ij}\Gamma_{j00} v^{0^2}$$
$$= -\frac{1}{2} g^{ij}\left(g_{j0,0} + g_{j0,0} - g_{00,j}\right) v^{0^2} = \frac{1}{2} g^{ij} g_{00,j} v^{0^2},$$

as the ordinary acceleration in the observational time x^0 is

$$\frac{\mathrm{d}^2x^i}{\mathrm{d}x^{0^2}} = \frac{\mathrm{d}^2x^i}{\mathrm{d}s^2}\left(\frac{\mathrm{d}s}{\mathrm{d}x^0}\right)^2 = \frac{\mathrm{d}v^i}{\mathrm{d}s} \cdot \frac{1}{v^{0^2}}.$$

We obtain

$$\frac{\mathrm{d}^2x^i}{\mathrm{d}x^{0^2}} = \frac{1}{2} g^{ij} g_{00,j},$$

which, for a diagonal metric tensor (3.128), is

$$\frac{\mathrm{d}^2x^i}{\mathrm{d}t^2} = -\frac{1}{2} c^2 g_{00,i} = -c^2 V_{,i}. \qquad \textbf{(3.130)}$$

For a radial motion in a system with spherical symmetry, this equation takes the form of Newton's gravitation law with the normalized gravitational potential (3.129) of the form

$$V = -m/r = -\frac{G}{c^2}\frac{M_G}{r}$$
$$= -\frac{6.6725\ 9\times10^{-11}\ \mathrm{m}^3\ \mathrm{s}^{-2}\ \mathrm{Kg}^{-1}}{299792458^2\ \mathrm{m}^2\ \mathrm{s}^{-2}}\frac{M_G}{r} = -7.4243\times10^{-28}\ \mathrm{m\,Kg}^{-1}\frac{M_G}{r}. \qquad \textbf{(3.131)}$$

For our planet with the mass $M_G = 5.972\times10^{24}$ Kg and the radius $r = 6.371\times10^6$ m , the normalized gravitational potential at its surface takes the value

$$V = -7.4243 \times 10^{-28} \mathrm{m\,Kg}^{-1} \frac{5.972 \times 10^{24} \mathrm{\,Kg}}{6.371 \times 10^{6} \mathrm{\,m}} = -6.9593 \times 10^{-10},$$

We obtain the acceleration (3.130)

$$a = -c^2 \frac{\partial V}{\partial r} = -\frac{mc^2}{r^2} = V \frac{c^2}{r}$$

$$= -6.9593 \times 10^{-10} \frac{\left(2.99792458 \times 10^{8}\right)^2 \mathrm{m}^2 \mathrm{s}^{-2}}{6.371 \times 10^{6} \mathrm{\,m}} = -9.8175 \mathrm{\,m\,s}^{-2},$$

in agreement with the well-known value of the gravitational acceleration at the surface of Earth. We consider an atom emitting light in a gravitational field with metric tensor (3.128), with a frequency $\omega = \frac{2\pi}{T} = \frac{2\pi c}{\Delta x^0}$. For the time-space interval invariant with the normalized gravitational potential (3.129), we obtain the expression:

$$\Delta s = \sqrt{g_{00} \Delta x^{0^2}} = \Delta x^0 \sqrt{1 + 2V} = \frac{2\pi c}{\omega} \sqrt{1 - \frac{2m}{r}},$$

which means that the frequency of the emitted light in a gravitational field with the potential (3.129) is

$$\omega = \frac{2\pi c}{\Delta s} \sqrt{1 - \frac{2m}{r}} \simeq \frac{2\pi c}{\Delta s} \left(1 - \frac{m}{r}\right). \tag{3.132}$$

We notice that the frequency of the emitted light decreases for smaller values of the distance r from the gravitational center – a redshift proportional to the gravitational potential where this light is emitted.

3.10. GRAVITATIONAL FIELD WITH SPHERICAL SYMMETRY – THE SCHWARZSCHILD SOLUTION

A flat physical space in spherical coordinates is described by the time-space interval

$$ds^2 = c^2 dt^2 - dr^2 - r^2 d\theta^2 - r^2 \sin^2 \theta d\varphi^2, \tag{3.133}$$

with the coordinates,

$$x^0 = ct,\ \ x^1 = r,\ \ x^2 = \theta,\ \ x^3 = \varphi,$$

and the metric elements:

$$\begin{array}{llll} g_{00} = 1 & g_{11} = -1 & g_{22} = -r^2 & g_{33} = -r^2 \sin^2\theta \\ g^{00} = 1 & g^{11} = -1 & g^{22} = -r^{-2} & g^{33} = -r^{-2}\sin^{-2}\theta. \end{array}$$

In a central gravitational field, such a physical space gets a curvature. For a constant gravitational field, which satisfies the conditions (3.125), such a curvature is described by a modification of the metric elements of the time-space interval (3.133). Unlike a weak gravitational field considered in the previous section, here we consider the general case of an arbitrary gravitational field. For this, we consider the modified time-space interval,

$$ds^2 = e^{2u(r)}c^2dt^2 - e^{2v(r)}dr^2 - r^2d\theta^2 - r^2\sin^2\theta d\varphi^2, \tag{3.134}$$

with the modified metric elements

$$\begin{array}{llll} g_{00} = e^{2u(r)} & g_{11} = -e^{2v(r)} & g_{22} = -r^2 & g_{33} = -r^2\sin^2\theta \\ g^{00} = e^{-2u(r)} & g^{11} = -e^{-2v(r)} & g^{22} = -r^{-2} & g^{33} = -r^{-2}\sin^{-2}\theta, \end{array} \tag{3.135}$$

depending on un the radius-dependent functions $u(r)$ and $v(r)$, which will be determined from Einstein's law of gravitation (3.126), with the explicit expression (3.124) of the Ricci tensor,

$$R_{\mu\nu} = -\Gamma^{\alpha}_{\mu\nu,\alpha} - \Gamma^{\beta}_{\mu\nu}\Gamma^{\alpha}_{\beta\alpha} + \Gamma^{\alpha}_{\mu\alpha,\nu} + \Gamma^{\beta}_{\mu\alpha}\Gamma^{\alpha}_{\beta\nu} = 0.$$

For a diagonalized form, we obtain the Ricci tensor elements:

$$R_{00} = -\Gamma^{\alpha}_{00,\alpha} - \Gamma^{\beta}_{00}\Gamma^{\alpha}_{\beta\alpha} + \Gamma^{\alpha}_{0\alpha,0} + \Gamma^{\beta}_{0\alpha}\Gamma^{\alpha}_{\beta 0} = 0 \tag{3.136}$$

$$R_{11} = -\Gamma^{\alpha}_{11,\alpha} - \Gamma^{\beta}_{11}\Gamma^{\alpha}_{\beta\alpha} + \Gamma^{\alpha}_{1\alpha,1} + \Gamma^{\beta}_{1\alpha}\Gamma^{\alpha}_{\beta 1} = 0 \tag{3.137}$$

$$R_{22} = -\Gamma^{\alpha}_{22,\alpha} - \Gamma^{\beta}_{22}\Gamma^{\alpha}_{\beta\alpha} + \Gamma^{\alpha}_{2\alpha,2} + \Gamma^{\beta}_{2\alpha}\Gamma^{\alpha}_{\beta 2} = 0 \tag{3.138}$$

$$R_{33} = -\Gamma^{\alpha}_{33,\alpha} - \Gamma^{\beta}_{33}\Gamma^{\alpha}_{\beta\alpha} + \Gamma^{\alpha}_{3\alpha,3} + \Gamma^{\beta}_{3\alpha}\Gamma^{\alpha}_{\beta 3} = 0 \,. \tag{3.139}$$

We calculate the Christoffel symbols of the form (3.98),

$$\Gamma^{\lambda}_{\nu\sigma} = g^{\lambda\mu}\Gamma_{\mu\nu\sigma} = \frac{1}{2} g^{\lambda\mu}\left(g_{\mu\nu,\sigma} + g_{\mu\sigma,\nu} - g_{\nu\sigma,\mu}\right),$$

according to the conditions (3.125). We obtain:

$$\Gamma^{0}_{00} = g^{0\mu}\Gamma_{\mu 00} = \frac{1}{2} g^{0\mu}\left(g_{\mu 0,0} + g_{\mu 0,0} - g_{00,\mu}\right) = \frac{1}{2} g^{00}\left(g_{00,0} + g_{00,0} - g_{00,0}\right) = 0$$

$$\begin{aligned} \Gamma^{1}_{00} &= g^{11}\Gamma_{100} = -e^{-2v(r)}\frac{1}{2}\left(\underbrace{g_{10,0}}_{0} + \underbrace{g_{10,0}}_{0} - g_{00,1}\right) \\ &= -e^{-2v(r)}\frac{1}{2}\left(-2u'(r)e^{2u(r)}\right) = u'(r)e^{2u(r)-2v(r)} \end{aligned} \tag{3.140}$$

$$\Gamma^{2}_{00} = g^{22}\Gamma_{200} = -r^{-2}\frac{1}{2}\left(g_{20,0} + g_{20,0} - g_{00,2}\right) = 0$$

$$\Gamma^{3}_{00} = g^{33}\Gamma_{300} = -r^{-2}\sin^{-2}\theta\frac{1}{2}\left(g_{30,0} + g_{30,0} - g_{00,3}\right) = 0$$

$$\begin{aligned} \Gamma^{0}_{01} &= g^{00}\Gamma_{001} = g^{00}\frac{1}{2}\left(g_{00,1} + \underbrace{g_{01,0}}_{0} - \underbrace{g_{01,0}}_{0}\right) \\ &= e^{-2u(r)}\frac{1}{2}2u'(r)e^{2u(r)} = u'(r) = \Gamma^{0}_{10} \end{aligned} \tag{3.141}$$

$$\Gamma^{1}_{01} = g^{11}\Gamma_{101} = e^{-2v(r)}\left(\underbrace{g_{10,1}}_{0} + \underbrace{g_{11,0}}_{0} - \underbrace{g_{01,1}}_{0}\right) = 0 = \Gamma^{1}_{10}$$

$$\Gamma^{2}_{01} = g^{22}\Gamma_{201} = -r^{-2}\left(\underbrace{g_{20,1}}_{0} + \underbrace{g_{21,0}}_{0} - \underbrace{g_{01,2}}_{0}\right) = 0 = \Gamma^{2}_{10}$$

$$\Gamma_{01}^{3}=g^{33}\Gamma_{301}=-r^{-2}\sin^{-2}\theta\left(\underset{0}{g_{30,1}}+\underset{0}{g_{31,0}}-\underset{0}{g_{01,3}}\right)=0=\Gamma_{10}^{3},$$

$$\Gamma_{02}^{0}=g^{00}\Gamma_{002}=e^{-2u(r)}\frac{1}{2}\left(\underset{0}{g_{00,2}}+\underset{0}{g_{02,0}}-\underset{0}{g_{02,0}}\right)=0=\Gamma_{20}^{0}$$

$$\Gamma_{02}^{1}=g^{11}\Gamma_{102}=-2e^{-2v(r)}\frac{1}{2}\left(\underset{0}{g_{10,2}}+\underset{0}{g_{12,0}}-\underset{0}{g_{02,1}}\right)=0=\Gamma_{20}^{1}$$

$$\Gamma_{02}^{2}=g^{22}\Gamma_{202}=-r^{-2}\frac{1}{2}\left(\underset{0}{g_{20,2}}+\underset{0}{g_{22,0}}-\underset{0}{g_{02,2}}\right)=0=\Gamma_{20}^{2}$$

$$\Gamma_{02}^{3}=g^{33}\Gamma_{302}=-r^{-2}\sin^{-2}\theta\frac{1}{2}\left(\underset{0}{g_{30,2}}+\underset{0}{g_{32,0}}-\underset{0}{g_{02,3}}\right)=0=\Gamma_{20}^{3}$$

$$\Gamma_{03}^{0}=g^{00}\Gamma_{003}=e^{-2u(r)}\frac{1}{2}\left(\underset{0}{g_{00,3}}+\underset{0}{g_{03,0}}-\underset{0}{g_{03,0}}\right)=0=\Gamma_{30}^{0}$$

$$\Gamma_{03}^{1}=g^{11}\Gamma_{103}=-e^{-2v(r)}\frac{1}{2}\left(\underset{0}{g_{10,3}}+\underset{0}{g_{13,0}}-\underset{0}{g_{30,1}}\right)=0=\Gamma_{30}^{1}$$

$$\Gamma_{03}^{2}=g^{22}\Gamma_{203}=-r^{-2}\frac{1}{2}\left(\underset{0}{g_{20,3}}+\underset{0}{g_{23,0}}-\underset{0}{g_{30,2}}\right)=0=\Gamma_{30}^{2}$$

$$\Gamma^{3}_{03} = g^{33}\Gamma_{303} = -r^{-2}\sin^{-2}\theta\frac{1}{2}\left(\underset{0}{g_{30,3}} + \underset{0}{g_{33,0}} - \underset{0}{g_{30,3}}\right) = 0 = \Gamma^{3}_{30}$$

$$\Gamma^{0}_{11} = g^{00}\Gamma_{011} = e^{-2u(r)}\frac{1}{2}\left(\underset{0}{g_{01,1}} + \underset{0}{g_{01,1}} - \underset{0}{g_{11,0}}\right) = 0$$

$$\begin{aligned}\Gamma^{1}_{11} &= g^{11}\Gamma_{111} = -e^{-2v(r)}\frac{1}{2}\left(g_{11,1} + \underline{g_{11,1}} - \underline{g_{11,1}}\right)\\ &= -e^{-2v(r)}\frac{1}{2}\left[-2v'(r)e^{2v(r)}\right] = v'(r)\end{aligned} \qquad \textbf{(3.142)}$$

$$\Gamma^{2}_{11} = g^{22}\Gamma_{211} = -r^{-2}\frac{1}{2}\left(\underset{0}{g_{21,1}} + \underset{0}{g_{21,1}} - \underset{0}{g_{11,2}}\right) = 0$$

$$\Gamma^{3}_{11} = g^{33}\Gamma_{311} = -r^{-2}\sin^{-2}\theta\left(\underset{0}{g_{31,1}} + \underset{0}{g_{31,1}} - \underset{0}{g_{11,3}}\right) = 0$$

$$\Gamma^{0}_{12} = g^{00}\Gamma_{012} = e^{-2u(r)}\frac{1}{2}\left(\underset{0}{g_{01,2}} + \underset{0}{g_{02,1}} - \underset{0}{g_{21,0}}\right) = 0 = \Gamma^{0}_{21}$$

$$\Gamma^{1}_{12} = g^{11}\Gamma_{112} = -e^{-2v(r)}\frac{1}{2}\left(\underset{0}{g_{11,2}} + \underset{0}{g_{12,1}} - \underset{0}{g_{21,1}}\right) = 0 = \Gamma^{1}_{21}$$

$$\begin{aligned}\Gamma^2_{12} &= g^{22}\Gamma_{212} = -r^{-2}\frac{1}{2}\left(\underbrace{g_{21,2}}_{0} + g_{22,1} - \underbrace{g_{21,2}}_{0}\right)\\ &= -r^{-2}\frac{1}{2}\left[-2r\right] = r^{-1} = \Gamma^2_{12}\end{aligned} \qquad \mathbf{(3.143)}$$

$$\Gamma^3_{12} = g^{33}\Gamma_{312} = -r^2\sin^{-2}\theta\frac{1}{2}\left(\underbrace{g_{31,2}}_{0} + \underbrace{g_{32,1}}_{0} - \underbrace{g_{21,3}}_{0}\right) = 0 = \Gamma^3_{21},$$

$$\Gamma^0_{13} = g^{00}\Gamma_{013} = e^{-2u(r)}\frac{1}{2}\left(\underbrace{g_{01,3}}_{0} + \underbrace{g_{03,1}}_{0} - \underbrace{g_{31,0}}_{0}\right) = 0 = \Gamma^0_{31}$$

$$\Gamma^1_{13} = g^{11}\Gamma_{113} = -e^{-2v(r)}\frac{1}{2}\left(\underbrace{g_{11,3}}_{0} + \underbrace{g_{13,1}}_{0} - \underbrace{g_{31,1}}_{0}\right) = 0 = \Gamma^1_{31}$$

$$\Gamma^2_{13} = g^{22}\Gamma_{213} = -r^{-2}\frac{1}{2}\left(\underbrace{g_{21,3}}_{0} + \underbrace{g_{23,1}}_{0} - \underbrace{g_{13,2}}_{0}\right) = 0 = \Gamma^2_{31}$$

$$\begin{aligned}\Gamma^3_{13} &= g^{33}\Gamma_{313} = -r^{-2}\sin^{-2}\theta\frac{1}{2}\left(\underbrace{g_{31,3}}_{0} + g_{33,1} - \underbrace{g_{13,3}}_{0}\right)\\ &= -r^{-2}\sin^{-2}\theta\frac{1}{2}\left[-2r\sin^2\theta\right] = r^{-1} = \Gamma^3_{31}\end{aligned} \qquad \mathbf{(3.144)}$$

$$\Gamma^0_{22} = g^{00}\Gamma_{022} = e^{-2u(r)}\frac{1}{2}\left(\underbrace{g_{02,2}}_{0} + \underbrace{g_{02,2}}_{0} - \underbrace{g_{22,0}}_{0}\right) = 0$$

$$\Gamma^1_{22} = g^{11}\Gamma_{122} = -e^{-2v(r)}\frac{1}{2}\left(\underbrace{g_{12,2}}_{0} + \underbrace{g_{12,2}}_{0} - g_{22,1}\right) = -e^{-2v(r)}\frac{1}{2}2r = -re^{-2v(r)} \qquad \mathbf{(3.145)}$$

$$\Gamma^2_{22} = g^{22}\Gamma_{222} = -r^{-2}\frac{1}{2}\left(\underbrace{g_{22,2}}_{0} + \underbrace{g_{22,2}}_{0} - \underbrace{g_{22,2}}_{0}\right) = 0$$

$$\Gamma^3_{22} = g^{33}\Gamma_{322} = -r^{-2}\sin^{-2}\theta\frac{1}{2}\left(\underset{0}{g_{32,2}} + \underset{0}{g_{32,2}} - \underset{0}{g_{22,3}}\right) = 0$$

$$\Gamma^0_{23} = g^{00}\Gamma_{023} = e^{-2u(r)}\frac{1}{2}\left(\underset{0}{g_{02,3}} + \underset{0}{g_{03,2}} - \underset{0}{g_{32,0}}\right) = 0 = \Gamma^0_{32}$$

$$\Gamma^1_{23} = g^{11}\Gamma_{123} = -e^{-2v(r)}\frac{1}{2}\left(\underset{0}{g_{12,3}} + \underset{0}{g_{13,2}} - \underset{0}{g_{32,1}}\right) = 0 = \Gamma^1_{32} \tag{3.146}$$

$$\Gamma^2_{23} = g^{22}\Gamma_{223} = -r^{-2}\frac{1}{2}\left(\underset{0}{g_{22,3}} + \underset{0}{g_{23,2}} - \underset{0}{g_{32,2}}\right) = 0 = \Gamma^2_{32}$$

$$\Gamma^3_{23} = g^{33}\Gamma_{323} = -r^{-2}\sin^{-2}\theta\frac{1}{2}\left(\underset{0}{g_{32,3}} + g_{33,2} - \underset{0}{g_{32,3}}\right)$$

$$= -r^{-2}\sin^{-2}\theta\frac{1}{2}\left(-2r^2\sin\theta\cos\theta\right) = \cot\theta = \Gamma^3_{32},$$

$$\Gamma^0_{33} = g^{00}\Gamma_{033} = e^{-2u(r)}\frac{1}{2}\left(\underset{0}{g_{03,3}} + \underset{0}{g_{03,3}} - \underset{0}{g_{33,0}}\right) = 0$$

$$\Gamma^1_{33} = g^{11}\Gamma_{133} = -e^{-2v(r)}\frac{1}{2}\left(\underset{0}{g_{13,3}} + \underset{0}{g_{13,3}} - g_{33,1}\right)$$

$$= -e^{-2v(r)}\frac{1}{2}\left(2r\sin^2\theta\right) = -r\sin^2\theta\, e^{-2v(r)} \tag{3.147}$$

$$\Gamma^2_{33} = g^{22}\Gamma_{233} = -r^{-2}\frac{1}{2}\left(\underset{0}{g_{23,3}} + \underset{0}{g_{23,3}} - g_{33,2}\right) \tag{3.148}$$

$$= -r^{-2}\frac{1}{2}\left(2r^2\sin\theta\cos\theta\right) = -\sin\theta\cos\theta = -\frac{1}{2}\sin(2\theta)$$

$$\Gamma^{3}_{33} = g^{33}\Gamma_{333} = -r^{-2}\sin^{-2}\theta\frac{1}{2}\left(\underbrace{g_{33,3}}_{0} + \underbrace{g_{33,3}}_{0} - \underbrace{g_{33,3}}_{0}\right) = 0.$$

With these Christoffel symbols, from (3.136) we obtain the equation

$$\begin{aligned}
R_{00} = & -\underbrace{\Gamma^{0}_{00,0}}_{0} - \Gamma^{1}_{00,1} - \underbrace{\Gamma^{2}_{00,2}}_{0} - \underbrace{\Gamma^{3}_{00,3}}_{0} + \underbrace{\Gamma^{0}_{00,0}}_{0} + \underbrace{\Gamma^{1}_{01,0}}_{0} + \underbrace{\Gamma^{2}_{02,0}}_{0} + \underbrace{\Gamma^{3}_{03,0}}_{0} \\
& -\underbrace{\Gamma^{0}_{00}}_{0}\Gamma^{0}_{00} - \underline{\Gamma^{1}_{00}}\Gamma^{0}_{10} - \underbrace{\Gamma^{2}_{00}}_{0}\Gamma^{0}_{20} - \underbrace{\Gamma^{3}_{00}}_{0}\Gamma^{0}_{30} - \underbrace{\Gamma^{0}_{00}}_{0}\Gamma^{1}_{01} - \underline{\Gamma^{1}_{00}}\Gamma^{1}_{11} - \underbrace{\Gamma^{2}_{00}}_{0}\Gamma^{1}_{21} - \underbrace{\Gamma^{3}_{00}}_{0}\Gamma^{1}_{31} \\
& -\underbrace{\Gamma^{0}_{00}}_{0}\Gamma^{2}_{02} - \underline{\Gamma^{1}_{00}}\Gamma^{2}_{12} - \underbrace{\Gamma^{2}_{00}}_{0}\Gamma^{2}_{22} - \underbrace{\Gamma^{3}_{00}}_{0}\Gamma^{2}_{32} - \underbrace{\Gamma^{0}_{00}}_{0}\Gamma^{3}_{03} - \underline{\Gamma^{1}_{00}}\Gamma^{3}_{13} - \underbrace{\Gamma^{2}_{00}}_{0}\Gamma^{3}_{23} - \underbrace{\Gamma^{3}_{00}}_{0}\Gamma^{3}_{33} \\
& +\underbrace{\Gamma^{0}_{00}}_{0}\Gamma^{0}_{00} + \underline{\underline{\Gamma^{1}_{00}\Gamma^{0}_{10}}} + \underbrace{\Gamma^{2}_{00}}_{0}\Gamma^{0}_{20} + \underbrace{\Gamma^{3}_{00}}_{0}\Gamma^{0}_{30} + \underline{\underline{\Gamma^{0}_{01}\Gamma^{1}_{00}}} + \underbrace{\Gamma^{1}_{01}}_{0}\Gamma^{1}_{10} + \underbrace{\Gamma^{2}_{01}}_{0}\Gamma^{1}_{20} + \underbrace{\Gamma^{3}_{01}}_{0}\Gamma^{1}_{30} \\
& +\underbrace{\Gamma^{0}_{02}}_{0}\Gamma^{2}_{00} + \underbrace{\Gamma^{1}_{02}}_{0}\Gamma^{2}_{10} + \underbrace{\Gamma^{2}_{02}}_{0}\Gamma^{2}_{20} + \underbrace{\Gamma^{3}_{02}}_{0}\Gamma^{2}_{30} + \underbrace{\Gamma^{0}_{03}}_{0}\Gamma^{3}_{00} + \underbrace{\Gamma^{1}_{03}}_{0}\Gamma^{3}_{10} + \underbrace{\Gamma^{2}_{03}}_{0}\Gamma^{1}_{20} + \underbrace{\Gamma^{3}_{03}}_{0}\Gamma^{3}_{30} \\
= & -\Gamma^{1}_{00,1} - \Gamma^{1}_{00}\left(\underline{\cancel{\Gamma^{0}_{10}}} + \Gamma^{1}_{11} + \Gamma^{2}_{12} + \Gamma^{3}_{13}\right) + \Gamma^{1}_{00}\underline{\cancel{\Gamma^{0}_{10}}} + \Gamma^{0}_{01}\Gamma^{1}_{00} \\
= & -\Gamma^{1}_{00,1} - \Gamma^{1}_{00}\left(\Gamma^{1}_{11} + \Gamma^{2}_{12} + \Gamma^{3}_{13}\right) + \Gamma^{0}_{01}\Gamma^{1}_{00} = 0,
\end{aligned}$$

with the expressions (3.140), (3.142), (3.143), (3.144), and (3.141):

$$\begin{aligned}
R_{00} &= -\Gamma^{1}_{00,1} - \Gamma^{1}_{00}\left(\Gamma^{1}_{11} + \Gamma^{2}_{12} + \Gamma^{3}_{13}\right) + \Gamma^{0}_{01}\Gamma^{1}_{00} \\
&= \left[u'(r)e^{2u(r)-2v(r)}\right]' - u'(r)e^{2u(r)-2v(r)}\left[v'(r) + r^{-1} + r^{-1}\right] + u'(r)u'(r)e^{2u(r)-2v(r)} \\
&= -\left[u''(r) + u'(r)\left(2\underline{\underline{u'(r)}} - 2\underline{v'(r)}\right)\right]e^{2u(r)-2v(r)} \\
&\quad - u'(r)\left(\underline{v'(r)} + 2r^{-1}\right)e^{2u(r)-2v(r)} + u'(r)\underline{\underline{u'(r)}}e^{2u(r)-2v(r)} \\
&= \left[-u''(r) - u'^{2}(r) + u'(r)v'(r) - 2r^{-1}u'(r)\right]e^{2u(r)-2v(r)} = 0.
\end{aligned} \qquad \textbf{(3.149)}$$

From (3.137), we obtain the equation

$$
\begin{aligned}
R_{11} &= -\underbrace{\Gamma^0_{11,0}}_{0} - \underline{\cancel{\Gamma^1_{11,1}}} - \underbrace{\Gamma^2_{11,2}}_{0} - \underbrace{\Gamma^3_{11,3}}_{0} + \Gamma^0_{10,1} + \underline{\cancel{\Gamma^1_{11,1}}} + \Gamma^2_{12,1} + \Gamma^3_{13,1} \\
&\quad -\underbrace{\Gamma^0_{11}}_{0}\Gamma^0_{00} - \Gamma^1_{11}\Gamma^0_{10} - \underbrace{\Gamma^2_{11}}_{0}\Gamma^0_{20} - \underbrace{\Gamma^3_{11}}_{0}\Gamma^0_{30} - \underbrace{\Gamma^0_{11}}_{0}\Gamma^1_{01} - \underline{\underline{\cancel{\Gamma^1_{11}\Gamma^1_{11}}}} - \underbrace{\Gamma^2_{11}}_{0}\Gamma^1_{21} - \underbrace{\Gamma^3_{11}}_{0}\Gamma^1_{31} \\
&\quad -\underbrace{\Gamma^0_{11}}_{0}\Gamma^2_{02} - \Gamma^1_{11}\Gamma^2_{12} - \underbrace{\Gamma^2_{11}}_{0}\Gamma^2_{22} - \underbrace{\Gamma^3_{11}}_{0}\Gamma^2_{32} - \underbrace{\Gamma^0_{11}}_{0}\Gamma^3_{03} - \Gamma^1_{11}\Gamma^3_{13} - \underbrace{\Gamma^2_{11}}_{0}\Gamma^3_{23} - \underbrace{\Gamma^3_{11}}_{0}\Gamma^3_{33} \\
&\quad +\Gamma^0_{10}\Gamma^0_{01} + \underbrace{\Gamma^1_{10}}_{0}\Gamma^0_{11} + \underbrace{\Gamma^2_{10}}_{0}\Gamma^0_{21} + \underbrace{\Gamma^3_{10}}_{0}\Gamma^0_{31} + \underbrace{\Gamma^0_{11}}_{0}\Gamma^1_{01} + \underline{\underline{\cancel{\Gamma^1_{11}\Gamma^1_{11}}}} + \underbrace{\Gamma^2_{11}}_{0}\Gamma^1_{21} + \underbrace{\Gamma^3_{11}}_{0}\Gamma^1_{31} \\
&\quad +\underbrace{\Gamma^0_{12}}_{0}\Gamma^2_{01} + \underbrace{\Gamma^1_{12}}_{0}\Gamma^2_{11} + \Gamma^2_{12}\Gamma^2_{21} + \underbrace{\Gamma^3_{12}}_{0}\Gamma^2_{31} + \underbrace{\Gamma^0_{13}}_{0}\Gamma^3_{01} + \underbrace{\Gamma^1_{13}}_{0}\Gamma^3_{11} + \underbrace{\Gamma^2_{13}}_{0}\Gamma^3_{21} + \Gamma^3_{13}\Gamma^3_{31} \\
&= -\Gamma^1_{11}\Gamma^0_{10} - \Gamma^1_{11}\Gamma^2_{12} - \Gamma^1_{11}\Gamma^3_{13} + \Gamma^0_{10,1} + \Gamma^2_{12,1} + \Gamma^3_{13,1} + \Gamma^0_{10}\Gamma^0_{01} + \Gamma^2_{12}\Gamma^2_{21} + \Gamma^3_{13}\Gamma^3_{31} \\
&= -\Gamma^1_{11}\left(\Gamma^0_{10} + \Gamma^2_{12} + \Gamma^3_{13}\right) + \Gamma^0_{10,1} + \Gamma^2_{12,1} + \Gamma^3_{13,1} + \Gamma^0_{10}\Gamma^0_{01} + \Gamma^2_{12}\Gamma^2_{21} + \Gamma^3_{13}\Gamma^3_{31} = 0,
\end{aligned}
$$

with the explicit form:

$$
\begin{aligned}
R_{11} &= -\Gamma^1_{11}\left(\Gamma^0_{10} + \Gamma^2_{12} + \Gamma^3_{13}\right) + \Gamma^0_{10,1} + \Gamma^2_{12,1} + \Gamma^3_{13,1} + \Gamma^0_{10}\Gamma^0_{01} + \Gamma^2_{12}\Gamma^2_{21} + \Gamma^3_{13}\Gamma^3_{31} \\
&= -v'(r)\left[u'(r) + r^{-1} + r^{-1}\right] + u''(r) - \underline{\cancel{r^{-2}}} - \underline{\cancel{r^{-2}}} + u'^2(r) + \underline{\cancel{r^{-2}}} + \underline{\cancel{r^{-2}}} \\
&= u''(r) + u'^2(r) - u'(r)v'(r) - 2r^{-1}v'(r) = 0.
\end{aligned}
\tag{3.150}
$$

From (3.138), we obtain the equation

$$
\begin{aligned}
R_{22} = & -\underbrace{\Gamma^0_{22,0}}_{0} - \Gamma^1_{22,1} - \underbrace{\Gamma^2_{22,2}}_{0} - \underbrace{\Gamma^3_{22,3}}_{0} + \underbrace{\Gamma^0_{20,2}}_{0} + \underbrace{\Gamma^1_{21,2}}_{0} + \underbrace{\Gamma^2_{22,2}}_{0} + \Gamma^3_{23,2} \\
& -\underbrace{\Gamma^0_{22}\Gamma^0_{00}}_{0} - \underline{\underline{\Gamma^1_{22}\Gamma^0_{10}}} - \underbrace{\Gamma^2_{22}\Gamma^0_{20}}_{0} - \underbrace{\Gamma^3_{22}\Gamma^0_{30}}_{0} - \underbrace{\Gamma^0_{22}\Gamma^1_{01}}_{0} - \underline{\underline{\Gamma^1_{22}\Gamma^1_{11}}} - \underbrace{\Gamma^2_{22}\Gamma^1_{21}}_{0} - \underbrace{\Gamma^3_{22}\Gamma^1_{31}}_{0} \\
& -\underbrace{\Gamma^0_{22}\Gamma^2_{02}}_{0} - \underline{\cancel{\Gamma^1_{22}\Gamma^2_{12}}} - \underbrace{\Gamma^2_{22}\Gamma^2_{22}}_{0} - \underbrace{\Gamma^3_{22}\Gamma^2_{32}}_{0} - \underbrace{\Gamma^0_{22}\Gamma^3_{03}}_{0} - \underline{\underline{\Gamma^1_{22}\Gamma^3_{13}}} - \underbrace{\Gamma^2_{22}\Gamma^3_{23}}_{0} - \underbrace{\Gamma^3_{22}\Gamma^3_{33}}_{0} \\
& +\underbrace{\Gamma^0_{20}\Gamma^0_{02}}_{0} + \underbrace{\Gamma^1_{20}\Gamma^0_{12}}_{0} + \underbrace{\Gamma^2_{20}\Gamma^0_{22}}_{0} + \underbrace{\Gamma^3_{20}\Gamma^0_{32}}_{0} + \underbrace{\Gamma^0_{21}\Gamma^1_{02}}_{0} + \underbrace{\Gamma^1_{21}\Gamma^1_{12}}_{0} + \underline{\Gamma^2_{21}\Gamma^1_{22}} + \underbrace{\Gamma^3_{21}\Gamma^1_{32}}_{0} \\
& +\underbrace{\Gamma^0_{22}\Gamma^2_{02}}_{0} + \underline{\cancel{\Gamma^1_{22}\Gamma^2_{12}}} + \underbrace{\Gamma^2_{22}\Gamma^2_{22}}_{0} + \underbrace{\Gamma^3_{22}\Gamma^2_{32}}_{0} + \underbrace{\Gamma^0_{23}\Gamma^3_{02}}_{0} + \underbrace{\Gamma^1_{23}\Gamma^3_{12}}_{0} + \underbrace{\Gamma^2_{23}\Gamma^3_{22}}_{0} + \Gamma^3_{23}\Gamma^3_{32} \\
= & -\Gamma^1_{22,1} + \Gamma^3_{23,2} - \Gamma^1_{22}\left(\Gamma^0_{10} + \Gamma^1_{11} + \Gamma^3_{13}\right) + \Gamma^1_{22}\Gamma^2_{12} + \Gamma^3_{23}\Gamma^3_{32} \\
= & -\Gamma^1_{22,1} + \Gamma^3_{23,2} - \Gamma^1_{22}\left(\Gamma^0_{10} + \Gamma^1_{11} - \Gamma^2_{12} + \Gamma^3_{13}\right) + \Gamma^3_{23}\Gamma^3_{32} = 0,
\end{aligned}
$$

with the explicit form:

$$
\begin{aligned}
R_{22} &= -\Gamma^1_{22,1} + \Gamma^3_{23,2} - \Gamma^1_{22}\left(\Gamma^0_{10} + \Gamma^1_{11} - \Gamma^2_{12} + \Gamma^3_{13}\right) + \Gamma^3_{23}\Gamma^3_{32} \\
&= -\left[-1 + \underline{\underline{2rv'(r)}}\right]e^{-2v(r)} - \underline{\sin^{-2}\theta} + re^{-2v(r)}\left[u'(r) + \underline{\underline{v'(r)}} - \underline{\cancel{r^{-1}}} + \underline{\cancel{r^{-1}}}\right] + \underline{\underline{\cot^2\theta}} \\
&= -1 + \left[1 + ru'(r) - rv'(r)\right]e^{-2v(r)} = 0.
\end{aligned}
\qquad \textbf{(3.151)}
$$

Finally, from (3.139) we obtain the equation

$$
\begin{aligned}
R_{33} = & -\underbrace{\Gamma^0_{33,0}}_{0} - \Gamma^1_{33,1} - \Gamma^2_{33,2} - \underbrace{\Gamma^3_{33,3}}_{0} + \underbrace{\Gamma^0_{30,3}}_{0} + \underbrace{\Gamma^1_{31,3}}_{0} + \underbrace{\Gamma^2_{32,3}}_{0} + \underbrace{\Gamma^3_{33,3}}_{0} \\
& -\underbrace{\Gamma^0_{33}\Gamma^0_{00}}_{0} - \underline{\Gamma^1_{33}\Gamma^0_{10}} - \underbrace{\Gamma^2_{33}\Gamma^0_{20}}_{0} - \underbrace{\Gamma^3_{33}\Gamma^0_{30}}_{0} - \underbrace{\Gamma^0_{33}\Gamma^1_{01}}_{0} - \underline{\Gamma^1_{33}\Gamma^1_{11}} - \underbrace{\Gamma^2_{33}\Gamma^1_{21}}_{0} - \underbrace{\Gamma^3_{33}\Gamma^1_{31}}_{0} \\
& -\underbrace{\Gamma^0_{33}\Gamma^2_{02}}_{0} - \underline{\Gamma^1_{33}\Gamma^2_{12}} - \underbrace{\Gamma^2_{33}\Gamma^2_{22}}_{0} - \underbrace{\Gamma^3_{33}\Gamma^2_{32}}_{0} - \underbrace{\Gamma^0_{33}\Gamma^3_{03}}_{0} - \underline{\cancel{\Gamma^1_{33}\Gamma^3_{13}}} - \underline{\underline{\cancel{\Gamma^2_{33}\Gamma^3_{23}}}} - \underbrace{\Gamma^3_{33}\Gamma^3_{33}}_{0} \\
& +\underbrace{\Gamma^0_{30}\Gamma^0_{03}}_{0} + \underbrace{\Gamma^1_{30}\Gamma^0_{13}}_{0} + \underbrace{\Gamma^2_{30}\Gamma^0_{23}}_{0} + \underbrace{\Gamma^3_{30}\Gamma^0_{33}}_{0} + \underbrace{\Gamma^0_{31}\Gamma^1_{03}}_{0} + \underbrace{\Gamma^1_{31}\Gamma^1_{13}}_{0} + \underbrace{\Gamma^2_{31}\Gamma^1_{23}}_{0} + \underline{\Gamma^3_{31}\Gamma^1_{33}} \\
& +\underbrace{\Gamma^0_{32}\Gamma^2_{03}}_{0} + \underbrace{\Gamma^1_{32}\Gamma^2_{13}}_{0} + \underbrace{\Gamma^2_{32}\Gamma^2_{23}}_{0} + \underline{\underline{\cancel{\Gamma^3_{32}\Gamma^2_{33}}}} + \underbrace{\Gamma^0_{33}\Gamma^3_{03}}_{0} + \underline{\cancel{\Gamma^1_{33}\Gamma^3_{13}}} + \underline{\underline{\Gamma^2_{33}\Gamma^3_{23}}} + \underbrace{\Gamma^3_{33}\Gamma^3_{33}}_{0} \\
= & -\Gamma^1_{33,1} - \Gamma^2_{33,2} - \Gamma^1_{33}\left(\Gamma^0_{10} + \Gamma^1_{11} + \Gamma^2_{12} - \Gamma^3_{31}\right) + \Gamma^2_{33}\Gamma^3_{23} = 0,
\end{aligned}
$$

with the explicit form:

$$\begin{aligned}
R_{33} &= -\Gamma^1_{33,1} - \Gamma^2_{33,2} - \Gamma^1_{33}\left(\Gamma^0_{10} + \Gamma^1_{11} + \Gamma^2_{12} - \Gamma^3_{31}\right) + \Gamma^2_{33}\Gamma^3_{23} \\
&= \sin^2\theta\left[1 - 2\underline{rv'(r)}\right]e^{-2v(r)} + \cos(2\theta) + \\
&\quad + r\sin^2\theta\, e^{-2v(r)}\left[u'(r) + \underline{v'(r)} + r^{-1} - r^{-1}\right] - \frac{1}{2}\sin(2\theta)\cot\theta \qquad \textbf{(3.152)} \\
&= \sin^2\theta\left[1 + ru'(r) - rv'(r)\right]e^{-2v(r)} + \underline{\cancel{\cos^2\theta} - \sin^2\theta - \cancel{\cos^2\theta}} \\
&= \sin^2\theta\left\{-1 + \left[1 + ru'(r) - rv'(r)\right]e^{-2v(r)}\right\} = \sin^2\theta R_{22} = 0.
\end{aligned}$$

Thus, the system of equations (3.149)-(3.152) is reduce to a system of three equations:

$$\begin{aligned}
u'(r) &= \frac{r}{2}\left[-u''(r) - u'^2(r) + u'(r)v'(r)\right] \\
v'(r) &= \frac{r}{2}\left[u''(r) + u'^2(r) - u'(r)v'(r)\right] \qquad \textbf{(3.153)} \\
&\left[1 + ru'(r) - rv'(r)\right]e^{-2v(r)} = 1.
\end{aligned}$$

From the first two equations we obtain

$$u'(r) + v'(r) = 0, \qquad \textbf{(3.154)}$$

with a solution

$$u(r) + v(r) = 0. \qquad \textbf{(3.155)}$$

From the third equation (3.153) with (3.154) and (3.155), we obtain the differential equations

$$\begin{aligned}
\left[1 + 2ru'(r)\right]e^{2u(r)} &= \left[re^{2u(r)}\right]' = 1 \\
\left[1 - 2rv'(r)\right]e^{-2v(r)} &= \left[re^{-2v(r)}\right]' = 1,
\end{aligned}$$

with the solutions

$$re^{2u(r)} = r - 2m$$

$$re^{-2v(r)} = r - 2m,$$

depending on the integration constant $2m$. With these expressions, we obtain the Schwarzschild metric elements (3.135):

$$g_{00} = 1 - \frac{2m}{r} \qquad g_{11} = -\left(1 - \frac{2m}{r}\right)^{-1} \qquad g_{22} = -r^2 \qquad g_{33} = -r^2 \sin^2\theta$$
$$g^{00} = \left(1 - \frac{2m}{r}\right)^{-1} \qquad g^{11} = -\left(1 - \frac{2m}{r}\right) \qquad g^{22} = -r^{-2} \qquad g^{33} = -r^{-2} \sin^{-2}\theta, \tag{3.156}$$

as the time-space interval (3.134) takes the explicit form:

$$\mathrm{d}s^2 = \left(1 - \frac{2m}{r}\right)c^2\mathrm{d}t^2 - \left(1 - \frac{2m}{r}\right)^{-1}\mathrm{d}r^2 - r^2\mathrm{d}\theta^2 - r^2\sin^2\theta\mathrm{d}\varphi^2. \tag{3.157}$$

3.11. QUANTUM PARTICLE IN A SPHERICAL GRAVITATIONAL FIELD AND BLACK HOLE

We consider a quantum particle in a spherical gravitational field, $x^1 = r,\ x^2 = \theta,\ x^3 = \varphi$, with a radial motion:

$$\frac{\mathrm{d}x^2}{\mathrm{d}s} = \frac{\mathrm{d}x^3}{\mathrm{d}s} = 0, \tag{3.158}$$

From the time geodesic equation (3.94) in the proper time,

$$\frac{\mathrm{d}^2x^0}{\mathrm{d}s^2} = -\Gamma^0_{01}\frac{\mathrm{d}x^0}{\mathrm{d}s}\frac{\mathrm{d}x^1}{\mathrm{d}s} - \Gamma^0_{10}\frac{\mathrm{d}x^1}{\mathrm{d}s}\frac{\mathrm{d}x^0}{\mathrm{d}s} = -2g^{00}\Gamma_{001}\frac{\mathrm{d}x^0}{\mathrm{d}s}\frac{\mathrm{d}x^1}{\mathrm{d}s}$$
$$= -g^{00}\left(g_{00,1} + g_{01,0} - g_{01,0}\right)\frac{\mathrm{d}x^0}{\mathrm{d}s}\frac{\mathrm{d}x^1}{\mathrm{d}s} = -g^{00}g_{00,1}\frac{\mathrm{d}x^0}{\mathrm{d}s}\frac{\mathrm{d}x^1}{\mathrm{d}s}$$
$$= -g^{00}\frac{\mathrm{d}g_{00}}{\mathrm{d}s}\frac{\mathrm{d}x^0}{\mathrm{d}s},$$

with the relation (3.15) between the covariant and contravariant metric tensors,

$$g^{00} = \frac{1}{g_{00}},$$

we obtain the differential equation

$$g_{00}\frac{\mathrm{d}^2 x^0}{\mathrm{d}s^2} + \frac{\mathrm{d}g_{00}}{\mathrm{d}s}\frac{\mathrm{d}x^0}{\mathrm{d}s} = 0,$$

with a solution of the form

$$g_{00}v^0 = k_0, \tag{3.159}$$

depending on the integration constant k_0. With this expression, from the fundamental equation (1.48), which in this case is

$$g_{00}v^{0^2} + g_{11}v^{1^2} = 1,$$

we obtain the radial velocity of a particle approaching the gravitational center:

$$v^1 = -\left(\frac{1-k_0 v^0}{g_{11}}\right)^{1/2} < 0.$$

With the Schwarzschild metric elements (3.156) and equation (3.159), this velocity takes the form

$$v^1 = -\left(k_0^2 - g_{00}\right)^{1/2}, \tag{3.160}$$

as the observation time variation in the proper time is

$$v^0 = \frac{k_0}{1-\frac{2m}{r}}. \tag{3.161}$$

From these expressions, we obtain the radial velocity in the observation time,

$$\frac{\mathrm{d}x^1}{\mathrm{d}x^0} = \frac{v^1}{v^0} = -\frac{1}{k_0}\left(k_0^2 + \frac{2m}{r} - 1\right)^{1/2}\left(1-\frac{2m}{r}\right).$$

This equation for a particle with a null velocity for $r \to \infty$, $k_0 = 1$, is of the form

$$\frac{dr}{dx^0} = -\left(1-\frac{r_0}{r}\right)\left(\frac{r_0}{r}\right)^{1/2}. \qquad \textbf{(3.162)}$$

This velocity is negative for a radius larger than the Schwarzschild radius which, according to (3.131), is proportional to the mass M_G generating this field:

$$r_0 = 2m = 2\frac{G}{c^2}M_G = 1.4849\times10^{-27}\,\mathrm{m\,Kg^{-1}}M_G. \qquad \textbf{(3.163)}$$

From the velocity (3.162), we obtain the acceleration in gravitational field,

$$\begin{aligned}
\frac{d^2r}{dx^{0^2}} &= -\frac{d}{dr}\left[\left(1-\frac{r_0}{r}\right)\left(\frac{r_0}{r}\right)^{1/2}\right]\frac{dr}{dx^0} \\
&= \left[\frac{r_0}{r^2}\left(\frac{r_0}{r}\right)^{1/2} + \left(1-\frac{r_0}{r}\right)\frac{1}{2}\left(\frac{r_0}{r}\right)^{-1/2}\left(-\frac{r_0}{r^2}\right)\right]\left(1-\frac{r_0}{r}\right)\left(\frac{r_0}{r}\right)^{1/2} \\
&= \left[\frac{r_0}{r^2}\frac{r_0}{r} + \left(1-\frac{r_0}{r}\right)\frac{1}{2}\left(-\frac{r_0}{r^2}\right)\right]\left(1-\frac{r_0}{r}\right) \\
&= -\left[-\frac{r_0}{r} + \left(1-\frac{r_0}{r}\right)\frac{1}{2}\right]\left(1-\frac{r_0}{r}\right)\frac{r_0}{r^2},
\end{aligned}$$

with the more convenient forms

$$\frac{d^2r}{dx^{0^2}} = -\frac{1}{2}\left[3\left(1-\frac{r_0}{r}\right)-2\right]\left(1-\frac{r_0}{r}\right)\frac{r_0}{r^2} = -\frac{1}{2}\left(1-\frac{3r_0}{r}\right)\left(1-\frac{r_0}{r}\right)\frac{r_0}{r^2}. \qquad \textbf{(3.164)}$$

From the metric elements (3.156), the time-space interval (157), the velocity (3.162), and the acceleration (3.164), we notice two singularities, for $r = 0$, and for the radius $r = r_0$.

If a large agglomeration of matter of mass M_G gravitationally concentrates inside a region with a radius smaller than the Schwarzschild radius (3.163), a black hole arise. From (3.162) we distinguish two spatial parts:

1) $r > r_0$ - the outside part, for $\frac{\mathrm{d}r}{\mathrm{d}x^0} < 0$, where any quantum particle approaches from infinity to the black hole, as a time-like region, with a positive time metric element $g_{00} = 1 - \frac{r_0}{r} > 0$, in the time-space interval (3.157),

$$\mathrm{d}s = \sqrt{\left(1 - \frac{r_0}{r}\right)c^2\mathrm{d}t^2 - \left(1 - \frac{r_0}{r}\right)^{-1}\mathrm{d}r^2 - r^2\mathrm{d}\theta^2 - r^2\sin^2\theta\mathrm{d}\varphi^2}\,, \qquad \textbf{(3.165)}$$

and

2) $r < r_0$ - the inside part of the black hole, for $\frac{\mathrm{d}r}{\mathrm{d}x^0} > 0$, where any particle moves towards the Schwarzschild boundary, as a space-like region, with a positive space metric element, $g_{11} = -\left(1 - \frac{r_0}{r}\right)^{-1} > 0$, as the time metric element is negative, $g_{00} = 1 - \frac{r_0}{r} < 0$. Any inside particle, moving towards the Schwarzschild boundary, is decelerated, $\frac{\mathrm{d}^2 r}{\mathrm{d}x^{0^2}} < 0$, so that it reaches this boundary with a null velocity, $\frac{\mathrm{d}r}{\mathrm{d}x^0} = 0$, null acceleration, $\frac{\mathrm{d}^2 r}{\mathrm{d}x^{0^2}} = 0$, *i.e.* in an infinite time.

From (3.164), we notice that the outside part of the black hole includes two regions:

1.1) $r > 3r_0$ - the far outside regions, for $\frac{\mathrm{d}^2 r}{\mathrm{d}x^{0^2}} < 0$, where any particle is attracted from infinity towards the black hole, and

1.2) $r_0 < r < 3r_0$ - the near outside region, for $\frac{\mathrm{d}^2 r}{\mathrm{d}x^{0^2}} > 0$, where any particle coming from infinity continues to approach the black hole, $\frac{\mathrm{d}r}{\mathrm{d}x^0} < 0$, but decelerated so that it reaches the boundary radius $r = r_0$ with a null velocity, $\frac{\mathrm{d}r}{\mathrm{d}x^0} = 0$, null acceleration, $\frac{\mathrm{d}^2 r}{\mathrm{d}x^{0^2}} = 0$, *i.e.* in an infinite time.

It is interesting to know what happens with a quantum particle at the two singularity points, $r = 0$ and $r = r_0$. We consider a quantum particle in the proper time τ, described by the wave functions

$$\begin{aligned} \psi\left(x^i,\tau\right) &= \frac{1}{\left(2\pi\hbar\right)^{3/2}} \int \varphi\left(P^j,\tau\right) e^{\frac{i}{\hbar}\left[P^j x^j - L\left(x^\alpha, v^\alpha\right)\tau\right]} \mathrm{d}^3 P \\ \varphi\left(P^j,\tau\right) &= \frac{1}{\left(2\pi\hbar\right)^{3/2}} \int \psi\left(x^i,\tau\right) e^{-\frac{i}{\hbar}\left[P^j x^j - L\left(x^\alpha, v^\alpha\right)\tau\right]} \mathrm{d}^3 x, \end{aligned} \tag{3.166}$$

with the Lagrangian

$$L\left(x^\alpha, v^\alpha\right) = -Mc^2 \sqrt{g_{\alpha\beta} v^\alpha v^\beta}\,, \tag{3.167}$$

in a system of coordinates

$$\left(x^\alpha\right) = \left(x^0 = ct, x^1, x^2, x^3\right) = \left(x^0 = ct, x^i\right), \tag{3.168}$$

with a differential time-space interval

$$\mathrm{d}s \equiv c\mathrm{d}\tau = \sqrt{g_{\alpha\beta}\mathrm{d}x^\alpha \mathrm{d}x^\beta} = \sqrt{g_{\alpha\beta} v^\alpha v^\beta}\,\mathrm{d}s\,, \tag{3.169}$$

and the conjugate space of the momentum

$$P^j = \frac{\partial L}{c\partial v^j} = -Mc\frac{\partial L}{\partial v^j}\sqrt{g_{\alpha\beta}v^\alpha v^\beta} = -Mc\frac{1}{2\sqrt{g_{\alpha\beta}v^\alpha v^\beta}}\frac{\partial L}{\partial v^j}\left(g_{j\beta}v^j v^\beta + g_{\alpha j}v^\alpha v^j\right)$$
$$= -Mc\frac{1}{2}\left(g_{j\beta}v^\beta + g_{\alpha j}v^\alpha\right) = -Mcg_{j\mu}v^\mu. \quad \textbf{(3.170)}$$

The particle dynamics is descried by the group velocities in the coordinate space,

$$\frac{\mathrm{d}}{\mathrm{d}\tau}x^j = \frac{\partial L}{\partial P^j} = -Mc^2\frac{\partial}{\partial\left(-Mcg_{j\mu}v^\mu\right)}\sqrt{g_{\alpha\beta}v^\alpha v^\beta}$$
$$= c\frac{1}{2\sqrt{g_{\alpha\beta}v^\alpha v^\beta}}\frac{\partial}{\partial\left(g_{j\mu}v^\mu\right)}\left(g_{j\mu}v^j v^\mu + g_{\mu j}v^\mu v^j\right) \quad \textbf{(3.171)}$$
$$= c\frac{1}{2}\frac{\partial}{\partial\left(g_{j\mu}v^\mu\right)}\left(2g_{j\mu}v^j v^\mu\right) = cv^j,$$

and in the momentum space:

$$\frac{\mathrm{d}}{\mathrm{d}\tau}P^j = \frac{\partial}{\partial x^j}L\left(x^\alpha, v^\alpha\right) = -Mc^2\frac{\partial}{\partial x^j}\sqrt{g_{\alpha\beta}v^\alpha v^\beta} =$$
$$= -Mc^2\frac{1}{2\sqrt{g_{\alpha\beta}v^\alpha v^\beta}}\frac{\partial}{\partial x^j}\left(g_{\alpha\beta}v^\alpha v^\beta\right) \quad \textbf{(3.172)}$$
$$= -\frac{1}{2}Mc^2 v^\alpha v^\beta g_{\alpha\beta,j}.$$

With the metric elements (3.156) for the radial motion, and the velocities (3.160) and (3.161) in the proper time, with the integration constant $k_0 = 1$ for a rest particle at infinity,

$$g_{00} = 1 - \frac{r_0}{r}, \qquad g_{11} = -\left(1 - \frac{r_0}{r}\right)^{-1}$$
$$v^0 = \frac{1}{1 - \frac{r_0}{r}}, \qquad v^1 = -\left(\frac{r_0}{r}\right)^{1/2}, \quad \textbf{(3.173)}$$

we obtain the Lagrangian function

$$L\left(r,\tau,v^0,v^1\right)=-Mc^2\sqrt{g_{00}v^{0^2}+g_{11}v^{1^2}}\,. \tag{3.174}$$

with a constant value

$$\begin{aligned} L\left(r,\tau,v^0,v^1\right)&=-Mc^2\sqrt{\left(1-\frac{2m}{r}\right)\frac{1}{\left(1-\frac{2m}{r}\right)^2}-\left(1-\frac{2m}{r}\right)^{-1}\frac{2m}{r}} \\ &=-Mc^2\sqrt{\frac{1}{1-\frac{2m}{r}}-\frac{1}{1-\frac{2m}{r}}\frac{2m}{r}}=-Mc^2. \end{aligned} \tag{3.175}$$

With this expression, the wave packets (3.166) describe an invariant distribution of matter in the coordinate space, determined only by the momentum dependent factor $\varphi\left(P^j,\tau\right)e^{\frac{i}{\hbar}P^jx^j}$, which, by this, is obtained as a constant function $\varphi\left(P^j,\tau\right)=\varphi\left(P^j\right)$. We obtain the momentum (3.170) of the form

$$\begin{aligned} P^1&=\frac{\partial L}{c\partial v^1}=-Mcg_{11}v^1 \\ &=\frac{Mc}{\frac{r_0}{r}-1}\left(\frac{r_0}{r}\right)^{1/2}=\frac{Mc}{\left(\frac{r_0}{r}\right)^{1/2}-\left(\frac{r}{r_0}\right)^{1/2}}. \end{aligned} \tag{3.176}$$

We notice that in a system with spherical symmetry, of a sufficiently large quantity of matter for becoming a black hole, a quantum particle in the center of this system, $r\approx 0$, has an approximately null momentum, $P\approx 0$. According to the wave function expression (3.166), this particle is crushed up in the momentum space, taking large dimensions in the coordinate space, $\Delta r\to\infty$. For a quantum particle at the Schwarzschild boundary, $r\simeq r_0$, from (3.176) we obtain an infinite momentum, according to (3.166) this particle being crushed up in the coordinate space, $\Delta r \simeq 0$. In the observational time, the two wave functions (3.166) take the form

$$\psi\left(x^{i},\tau\right)=\frac{1}{\left(2\pi\hbar\right)^{3/2}}\int\varphi\left(P^{j},\tau\right)e^{\frac{\mathrm{i}}{\hbar}\left[P^{j}x^{j}-L\left(x^{\alpha},\dot{x}^{\alpha}\right)t\right]}\mathrm{d}^{3}P \tag{3.177}$$

$$\varphi\left(P^{j},\tau\right)=\frac{1}{\left(2\pi\hbar\right)^{3/2}}\int\psi\left(x^{i},\tau\right)e^{-\frac{\mathrm{i}}{\hbar}\left[P^{j}x^{j}-L\left(x^{\alpha},\dot{x}^{\alpha}\right)t\right]}\mathrm{d}^{3}x, \tag{3.178}$$

with the Lagrangian

$$L\left(x^{1},\dot{x}^{1}\right)=-Mc^{2}\sqrt{g_{00}\dot{x}^{0^{2}}+g_{11}\dot{x}^{1^{2}}}=-Mc^{2}\sqrt{g_{00}+g_{11}\dot{x}^{1^{2}}} \tag{3.179}$$

and the velocity (3.162),

$$\dot{x}^{1}=\frac{\mathrm{d}r}{\mathrm{d}x^{0}}=-\left(1-\frac{r_{0}}{r}\right)\left(\frac{r_{0}}{r}\right)^{1/2}. \tag{3.180}$$

We obtain the Lagrangian

$$\begin{aligned}L\left(x^{1},\dot{x}^{1}\right)&=-Mc^{2}\sqrt{g_{00}+g_{11}\dot{x}^{1^{2}}}\\&=-Mc^{2}\sqrt{1-\frac{r_{0}}{r}-\left(1-\frac{r_{0}}{r}\right)^{-1}\left(1-\frac{r_{0}}{r}\right)^{2}\frac{r_{0}}{r}}\\&=-Mc^{2}\left(1-\frac{r_{0}}{r}\right),\end{aligned} \tag{3.181}$$

and the momentum

$$P^1=\frac{\partial L}{c\partial \dot{x}^1}=-Mc\frac{\partial}{\partial \dot{x}^1}\sqrt{g_{00}+g_{11}\dot{x}^{1^2}}=-Mc\frac{g_{11}\dot{x}^1}{\sqrt{g_{00}+g_{11}\dot{x}^{1^2}}}$$

$$=-Mc\frac{\left(1-\frac{r_0}{r}\right)^{-1}\left(1-\frac{r_0}{r}\right)\left(\frac{r_0}{r}\right)^{1/2}}{\left(1-\frac{r_0}{r}\right)}=\frac{Mc}{\frac{r_0}{r}-1}\left(\frac{r_0}{r}\right)^{1/2} \qquad \textbf{(3.182)}$$

$$=\frac{Mc}{\left(\frac{r_0}{r}\right)^{1/2}-\left(\frac{r}{r_0}\right)^{1/2}},$$

of the form (3.176), as a function of the spatial coordinate. From the group velocity of the wave function (3.178) and the Lagrangian (3.181), we obtain the dynamic equation

$$\frac{\mathrm{d}}{\mathrm{d}t}P^1=\frac{\partial}{\partial x^1}L\left(x^1,\dot{x}^1\right)=-Mc^2\frac{\partial}{\partial x^1}\sqrt{g_{00}+g_{11}\dot{x}^{1^2}}$$

$$=-Mc^2\frac{1}{2\sqrt{g_{00}+g_{11}\dot{x}^{1^2}}}\left(\frac{\partial g_{00}}{\partial x^1}+\frac{\partial g_{11}}{\partial x^1}\dot{x}^{1^2}\right)$$

$$=-Mc^2\frac{1}{2\left(1-\frac{r_0}{r}\right)}\left[\frac{\partial}{\partial r}\left(1-\frac{r_0}{r}\right)-\frac{\partial}{\partial r}\left(1-\frac{r_0}{r}\right)^{-1}\left(1-\frac{r_0}{r}\right)^2\frac{r_0}{r}\right]$$

$$=-\frac{Mc^2}{2\left(1-\frac{r_0}{r}\right)}\left[\frac{r_0}{r^2}+\cancel{\left(1-\frac{r_0}{r}\right)^{-2}}\frac{r_0}{r^2}\cancel{\left(1-\frac{r_0}{r}\right)^2}\frac{r_0}{r}\right]$$

$$=-\frac{1}{2}\left(1+\frac{r_0}{r}\right)Mc^2\frac{1}{1-\frac{r_0}{r}}\frac{r_0}{r^2}=\frac{1}{2}\left(1+\frac{r_0}{r}\right)\frac{1}{1-\frac{r}{r_0}}\frac{Mc^2}{r}. \qquad \textbf{(3.183)}$$

From this expression, we notice that at the moment of the black hole formation, the central particles are thrown up with a huge momentum variation $\frac{Mc^2 r_0}{r^2}$, and velocity (3.180). At the Schwarzschild boundary, $r \to r_0$, the momentum variation has a singularity with a change of sign, $\frac{\mathrm{d}}{\mathrm{d}t} P^1 = \pm\infty$, as the velocity (3.180) is null – the particle is reflected at this boundary.

A black hole arise when a sufficiently large mass M_G with the mean density ρ_G is accumulated inside a sphere with the Schwarzschild radius r_0. According to (3.163), these quantities satisfy the relation

$$r_0 = 1.4849 \times 10^{-27}\,\mathrm{m\,Kg}^{-1}\,\rho_G \frac{4\pi}{3} r_0^{\,3},$$

which leads to the formation condition of a black hole:

$$\rho_G > 1.6078 \times 10^{26} \frac{\mathrm{Kg}}{\mathrm{m}} \frac{1}{r_0^{\,2}}. \quad \textbf{(3.184)}$$

For a black hole with the radius of Sun, $r_0 = 696\ 340\,\mathrm{Km} = 6.9634 \times 10^{8}\,\mathrm{m}$, we find the condition

$$\rho_G > 3.3158 \times 10^{8} \frac{\mathrm{Kg}}{\mathrm{m}^3} = 3.3158 \times 10^{5} \frac{\mathrm{Kg}}{\mathrm{dm}^3}, \quad \textbf{(3.185)}$$

as the density of Sun with the mass $1.9885 \times 10^{30}\,\mathrm{Kg}$ is

$$\rho_G^{\mathrm{Sun}} = 1.4060 \times 10^{3} \frac{\mathrm{Kg}}{\mathrm{m}^3} = 1.4060 \frac{\mathrm{Kg}}{\mathrm{dm}^3}, \quad \textbf{(3.186)}$$

i.e. much smaller than the minimum value (3.185) for a black hole.

3.12. EINSTEIN'S LAW OF THE GRAVITATIONAL FIELD AND THE BLACK HOLE DYNAMICS

In section 3.8 we obtained Bianci's field equation (3.122),

$$\left(R^{\mu\nu}-\frac{1}{2}g^{\mu\nu}R\right)_{:\nu}=0. \quad \textbf{(3.187)}$$

In section 3.10, we considered the simpler solution

$$R_{\mu\nu}=0$$

and, for a system with spherical symmetry we obtained the Schwarzschild metric tensor:

$$\begin{aligned} &g_{00}=1-\frac{2m}{r} \quad g_{11}=-\left(1-\frac{2m}{r}\right)^{-1} \quad g_{22}=-r^2 \quad g_{33}=-r^2\sin^2\theta \\ &g^{00}=\left(1-\frac{2m}{r}\right)^{-1} \quad g^{11}=-\left(1-\frac{2m}{r}\right) \quad g^{22}=-r^{-2} \quad g^{33}=-r^{-2}\sin^{-2}\theta. \end{aligned} \quad \textbf{(3.188)}$$

In section 3.3, we found that the matter conservation is described by a null covariant divergence of the matter flow four-vector:

$$\left(\rho_G v^\mu\right)_{:\mu}=0.$$

At the same time, in section 3.6 we showed that the dynamics of a particle in a gravitational field satisfies the geodesic equation of a null covariant acceleration:

$$v^\mu{}_{:\nu}v^\nu=0.$$

From these two equations, we find that the tensor

$$T^{\mu\nu}=\rho_G v^\mu v^\nu \quad \textbf{(3.189)}$$

has a null covariant divergence:

$$T^{\mu\nu}{}_{:\nu}=\left(\rho_G v^\mu v^\nu\right)_{:\nu}=\rho_G v^\mu{}_{:\nu}v^\nu+v^\mu\left(\rho_G v^\nu\right)_{:\nu}=0. \quad \textbf{(3.190)}$$

From (3.187) and (3.189)-(3.190), we obtain Einstein's law of the gravitational field as the time-space curvature created by matter:

$$R^{\mu\nu}-\frac{1}{2}g^{\mu\nu}R=k\rho v^\mu v^\nu, \quad \textbf{(3.191)}$$

where ρ is a normalized density with a coefficient k, which will be determined from the corresponding Newtonian expression for a weak gravitational field in a system with spherical symmetry. By multiplying this equation with the covariant metric tensor,

$$g_{\mu\nu}R^{\mu\nu} - \frac{1}{2}g_{\mu\nu}g^{\mu\nu}R = k\rho g_{\mu\nu}v^{\mu}v^{\nu}$$
$$R - 2R = k\rho g_{\mu\nu}v^{\mu}v^{\nu},$$

with the fundamental equation (1.48),

$$g_{\mu\nu}v^{\mu}v^{\nu} = 1, \tag{3.192}$$

we obtain

$$R = -k\rho. \tag{3.193}$$

With this expression, equation (3.191) takes the form

$$R^{\mu\nu} = k\rho\left(v^{\mu}v^{\nu} - \frac{1}{2}g^{\mu\nu}\right). \tag{3.194}$$

which, by lowering the indices, takes the similar form

$$R_{\mu\nu} = k\rho\left(v_{\mu}v_{\nu} - \frac{1}{2}g_{\mu\nu}\right). \tag{3.195}$$

Thus we obtain an equation for the covariant Ricci tensor,

$$R_{\mu\nu} = R^{\rho}_{\mu\nu\rho} = g^{\rho\sigma}R_{\sigma\mu\nu\rho} = k\rho\left(v_{\mu}v_{\nu} - \frac{1}{2}g_{\mu\nu}\right).$$

For a sufficiently weak gravitational field, with the first-order expression (3.104) with (3.106),

$$R_{\mu\nu\rho\sigma} = \frac{1}{2}\left(g_{\mu\sigma,\nu\rho} - g_{\nu\sigma,\mu\rho} + g_{\nu\rho,\mu\sigma} - g_{\mu\rho,\nu\sigma}\right), \tag{3.196}$$

we obtain

$$\frac{1}{2}g^{\sigma\rho}\left(g_{\sigma\rho,\mu\nu}-g_{\sigma\nu,\mu\rho}+g_{\mu\nu,\sigma\rho}-g_{\mu\rho,\sigma\nu}\right)=k\rho\left(v_{\mu}v_{\nu}-\frac{1}{2}g_{\mu\nu}\right).$$

This equation for $\mu=\nu=0$,

$$\frac{1}{2}g^{\sigma\rho}\left(g_{\sigma\rho,00}-g_{\sigma\nu,0\rho}+g_{00,\sigma\rho}-g_{0\rho,\sigma0}\right)=k\rho\left(v_{0}v_{0}-\frac{1}{2}g_{00}\right),$$

with the conditions (3.125) for a constant gravitational field, the metric elements (3.128) for a weak gravitational field, $g_{00}=1+2V\simeq 1,\ g^{11}=-1$, and a small velocity, $v_0\simeq 1,\ \ v_m\simeq 0$, is

$$\nabla^2 V=-\frac{1}{2}k\rho .$$

By integration over the volume V_S of a sphere S,

$$\int_{V_S}\nabla^2 V\mathrm{d}^3x=\oint_S \nabla V\mathrm{d}^2x=-\frac{1}{2}k\int_{V_S}\rho\mathrm{d}^3x,$$

we obtain

$$4\pi r^2\nabla V\frac{\vec{r}}{r}=-\frac{1}{2}k\int_{V_S}\rho\mathrm{d}^3x .$$

With the Newtonian potential (3.129),

$$V=-\frac{m}{r},$$

this equation takes the form

$$4\pi r^2\frac{m}{r^2}\frac{\vec{r}}{r}\frac{\vec{r}}{r}=-\frac{1}{2}k\int_{V_S}\rho\mathrm{d}^3x$$

similar to (3.163)

$$m=\frac{G}{c^2}M_G=\frac{G}{c^2}\int_{V_S}\rho_G\mathrm{d}^3x=-\frac{k}{8\pi}\int_{V_S}\rho\mathrm{d}^3x .$$

With

$$k = -8\pi ,$$

we obtain the normalized density ρ as a function of the mass density ρ_G:

$$\rho = \frac{G}{c^2}\rho_G \ \left[\mathrm{m}^{-2}\right], \qquad \textbf{(3.197)}$$

as Einstein's gravitation law (3.191) takes the explicit form

$$R^{\mu\nu} - \frac{1}{2}g^{\mu\nu}R = -8\pi\rho v^{\mu}v^{\nu} . \qquad \textbf{(3.198)}$$

With this expression, the gravitational field equation (3.193) with (3.197) takes the explicit form

$$R = 8\pi\frac{G}{c^2}\rho_G . \qquad \textbf{(3.199)}$$

For a black hole with the total mass M_G, from this equation we obtain the dynamic equation

$$\int_0^\infty R4\pi r^2 \mathrm{d}r = 8\pi\frac{G}{c^2}\int_0^\infty \rho_G = 8\pi\frac{G}{c^2}M_G = 8\pi\, m$$

which is

$$\frac{1}{2}\int_0^\infty Rr^2\mathrm{d}r = m . \qquad \textbf{(3.200)}$$

We consider the expression (3.121) of the time-space total curvature in spherical coordinates

$$R = g^{\mu\nu}R_{\mu\nu} = g^{00}R_{00} + g^{11}R_{11} , \qquad \textbf{(3.201)}$$

and calculate the curvature matrix elements in the first-order approximation, with the expression (3.196),

$$R_{\mu\nu} = R^{\alpha}_{\mu\nu\alpha} = g^{\alpha\beta} R_{\beta\mu\nu\alpha} = \frac{1}{2} g^{\alpha\beta} \left(g_{\beta\alpha,\mu\nu} - g_{\beta\nu,\mu\alpha} + g_{\mu\nu,\beta\alpha} - g_{\mu\alpha,\beta\nu} \right). \qquad \textbf{(3.202)}$$

We obtain

$$\begin{aligned} R_{00} &= \frac{1}{2} g^{\alpha\beta} \left(g_{\beta\alpha,00} - g_{\beta 0,0\alpha} + g_{00,\beta\alpha} - g_{0\alpha,\beta 0} \right) \\ &= \frac{1}{2} \left(g^{\alpha\alpha} g_{\alpha\alpha,00} - g^{\alpha\alpha} g_{\alpha 0,0\alpha} + g^{\alpha\alpha} g_{00,\alpha\alpha} - g^{\alpha\alpha} g_{0\alpha,\alpha 0} \right) \\ &= \frac{1}{2} \left(g^{\alpha\alpha} g_{\alpha\alpha,00} - g^{00} g_{00,00} + g^{\alpha\alpha} g_{00,\alpha\alpha} - g^{00} g_{00,00} \right) \\ &= \frac{1}{2} \left(\underline{g^{00} g_{00,00}} + g^{11} g_{11,00} - \underline{g^{00} g_{00,00}} + \underline{\underline{g^{00} g_{00,00}}} + g^{11} g_{00,11} - \underline{\underline{g^{00} g_{00,00}}} \right) \\ &= \frac{1}{2} g^{11} \left(g_{11,00} + g_{00,11} \right), \end{aligned}$$

and

$$\begin{aligned} R_{11} &= \frac{1}{2} g^{\alpha\beta} \left(g_{\beta\alpha,11} - g_{\beta 1,1\alpha} + g_{11,\beta\alpha} - g_{1\alpha,\beta 1} \right) \\ &= \frac{1}{2} \left(g^{\alpha\alpha} g_{\alpha\alpha,11} - g^{11} g_{11,11} + g^{\alpha\alpha} g_{11,\alpha\alpha} - g^{11} g_{11,11} \right) \\ &= \frac{1}{2} \left(g^{00} g_{00,11} + \underline{g^{11} g_{11,11}} - \underline{g^{11} g_{11,11}} + g^{00} g_{11,00} + \underline{\underline{g^{11} g_{11,11}}} - \underline{\underline{g^{11} g_{11,11}}} \right) \\ &= \frac{1}{2} g^{00} \left(g_{00,11} + g_{11,00} \right). \end{aligned}$$

With the Schwarzschild matrix elements

$$g_{00} = 1 - \frac{2m}{r}, \quad g^{00} = \left(1 - \frac{2m}{r}\right)^{-1}, \quad g_{11} = -\left(1 - \frac{2m}{r}\right)^{-1}, \quad g^{11} = -\left(1 - \frac{2m}{r}\right), \qquad \textbf{(3.203)}$$

for the total curvature, we obtain the expression

$$\begin{aligned} R &= g^{00} R_{00} + g^{11} R_{11} = g^{00} \frac{1}{2} g^{11} \left(\underline{g_{11,00}} + \underline{\underline{g_{00,11}}} \right) + g^{11} \frac{1}{2} g^{00} \left(\underline{\underline{g_{00,11}}} + \underline{g_{11,00}} \right) \\ &= -\left(g_{11,00} + g_{00,11} \right). \end{aligned} \qquad \textbf{(3.204)}$$

For the second derivative with radius of the time metric element, from (3.203) we obtain

$$g_{00,11} = -2\frac{2m}{r^3}. \quad \textbf{(3.205)}$$

For (3.204), we calculate the second derivative with time of the radius metric element by using the expressions (3.162) and (3.164):

$$\begin{aligned}\frac{\mathrm{d}r}{\mathrm{d}x^0} &= -\left(1-\frac{2m}{r}\right)\left(\frac{2m}{r}\right)^{1/2}\\ \frac{\mathrm{d}^2 r}{\mathrm{d}x^{0^2}} &= -\left[3\left(1-\frac{2m}{r}\right)-2\right]\left(1-\frac{2m}{r}\right)\frac{m}{r^2}.\end{aligned} \quad \textbf{(3.206)}$$

From the time derivative of the radius metric element

$$g_{11,0} = \left(1-\frac{2m}{r}\right)^{-2}\left(\frac{2m}{r^2}\right)\frac{\mathrm{d}r}{\mathrm{d}x^0},$$

we obtain the second derivative:

$$\begin{aligned} g_{11,00} &= \left[-2\left(1-\frac{2m}{r}\right)^{-3}\left(\frac{2m}{r^2}\right)^2 - 2\left(1-\frac{2m}{r}\right)^{-2}\left(\frac{2m}{r^3}\right)\right]\left(\frac{\mathrm{d}r}{\mathrm{d}x^0}\right)^2 + \left(1-\frac{2m}{r}\right)^{-2}\left(\frac{2m}{r^2}\right)\frac{\mathrm{d}^2 r}{\mathrm{d}x^{0^2}}\\ &= -2\left[\left(1-\frac{2m}{r}\right)^{-1}\frac{2m}{r}+1\right]\left(1-\frac{2m}{r}\right)^{-2}\frac{2m}{r^3}\left(\frac{\mathrm{d}r}{\mathrm{d}x^0}\right)^2 + \left(1-\frac{2m}{r}\right)^{-2}\left(\frac{2m}{r^2}\right)\frac{\mathrm{d}^2 r}{\mathrm{d}x^{0^2}}\\ &= -2\left(1-\frac{2m}{r}\right)^{-3}\frac{2m}{r^3}\left(\frac{\mathrm{d}r}{\mathrm{d}x^0}\right)^2 + \left(1-\frac{2m}{r}\right)^{-2}\left(\frac{2m}{r^2}\right)\frac{\mathrm{d}^2 r}{\mathrm{d}x^{0^2}}.\end{aligned}$$

With (3.206), this expression takes the form

$$g_{11,00}=-2\left(1-\frac{2m}{r}\right)^{-3}\frac{2m}{r^3}\left(1-\frac{2m}{r}\right)^2\left(\frac{2m}{r}\right)-\left(1-\frac{2m}{r}\right)^{-2}\left(\frac{2m}{r^2}\right)\left[3\left(1-\frac{2m}{r}\right)-2\right]\left(1-\frac{2m}{r}\right)\frac{m}{r^2}$$
$$=-2\left\{4+\left[3\left(1-\frac{2m}{r}\right)-2\right]\right\}\left(1-\frac{2m}{r}\right)^{-1}\left(\frac{m}{r^2}\right)^2$$
$$=-2\left(3+\frac{2}{1-\frac{2m}{r}}\right)\left(\frac{m}{r^2}\right)^2. \tag{3.207}$$

From (3.204) with (3.205) and (3.207), we obtain the total curvature:

$$R=\left[2+\frac{m}{r}\left(3+\frac{2}{1-\frac{2m}{r}}\right)\right]\frac{2m}{r^3}. \tag{3.208}$$

With this expression, we consider the dynamic equation (3.200) as an integral between two finite limits of the matter distribution,

$$\int_{r_1}^{r_2}\left[1+\frac{m}{r}\left(\frac{3}{2}+\frac{1}{1-\frac{2m}{r}}\right)\right]\frac{2}{r}\mathrm{d}r=1, \tag{3.209}$$

with the form

$$\int_{r_1}^{r_2}\left(\frac{2r-2m}{r^2-2mr}+\frac{3m}{r^2}\right)\mathrm{d}r=1,$$

which can be analytically integrated:

$$\ln\left|r(r-2m)\right|\Big|_{r_1}^{r_2}-\frac{3m}{r}\Big|_{r_1}^{r_2}=1. \tag{3.210}$$

We consider the integration limits normalized to the black hole radius $r_0=2m$,

$$x_1=\frac{r_1}{2m}=\frac{r_1}{r_0},\qquad x_2=\frac{r_2}{2m}=\frac{r_2}{r_0}, \tag{3.211}$$

and the parameter

$$\eta = \frac{x_2\left(x_2 - 1\right)}{x_1\left(1 - x_1\right)}. \tag{3.212}$$

With these notations, the upper limit of the matter distribution as a function of the lower limit is

$$x_2 = \frac{1}{2} + \sqrt{\frac{1}{4} + \eta\left(x_1 - x_1^{\ 2}\right)}, \tag{3.213}$$

with equation (3.210) which is

$$\ln\eta + \frac{3}{2}\left(\frac{1}{x_1} - \frac{1}{x_2}\right) = 1. \tag{3.214}$$

These equations describe the evolution of the matter from the initial state $x_1 = 0,\ x_2 = 1$, of the matter entirely comprised in the sphere with the Schwarzschild radius $r_0 = 2m$, to the final state $x_1 = 1,\ x_2 = 1$ of the matter placed in the neighborhood of the Schwarzschild radius $r_0 = 2m$:

$$\begin{cases} x_1 = 0 \quad \rightarrow \quad x_2 = 1 \\ x_1 = \dfrac{1}{2} \quad \rightarrow \ln\eta = -\dfrac{2\sqrt{1+\eta} - 1}{\sqrt{1+\eta} + 1}, \quad \eta = 0.52, \quad x_2 = \dfrac{1}{2}\left(1 + \sqrt{1+\eta}\right) = 1.116 \\ x_1 = 1 \quad \rightarrow \quad x_2 = 1. \end{cases} \tag{3.215}$$

We notice that between the initial time $x^0{}_1 = 0$, to the final time $x^0{}_2 = \infty$, the matter accumulates in the neighborhood of the Schwarzschild radius, in the intermediate times exceeding a little this radius.

From the exact expression (3.206) of a particle velocity in an arbitrary central gravitational field, we obtain the differential equation of the time variation as a function of the radial displacement:

$$\mathrm{d}x^0 = \left(\frac{2m}{r} - 1\right)^{-1}\left(\frac{2m}{r}\right)^{-1/2}\mathrm{d}r,$$

which we integrate with the notations

$$x=\frac{r}{2m},\qquad dr=2m\mathrm{d}x,\qquad x=y^2,\qquad \mathrm{d}x=2y\mathrm{d}y\,. \tag{3.216}$$

We obtain

$$\mathrm{d}x^0=2m\frac{x^{3/2}}{1-x}\mathrm{d}x=4m\left(\frac{1}{1-y^2}-1-y^2\right)\mathrm{d}y\,,$$

which, by integration, leads to the evolution equation

$$\begin{aligned} x^0(r)&=4m\left\{\frac{1}{2}\left[\ln(1+y)-\ln(1-y)\right]-y-\frac{1}{3}y^3\right\}\\ &=4m\left\{\frac{1}{2}\left[\ln\left(1+\sqrt{\frac{r}{2m}}\right)-\ln\left(1-\sqrt{\frac{r}{2m}}\right)\right]-\sqrt{\frac{r}{2m}}-\frac{1}{3}\left(\frac{r}{2m}\right)^{3/2}\right\}. \end{aligned} \tag{3.217}$$

We notice this equation describes an evolution from $\left(r=0,\ x^0=0\right)$ to $\left(r=2m=r_0,\ x^0=\infty\right)$. With the series expansion of the logarithm,

$$\ln(1+y)=y-\frac{1}{2}y^2+\frac{1}{3}y^3+\ldots,$$

we obtain the time as a function of radius as power series expansion

$$\begin{aligned} x^0(r)=ct(r)&=4m\left\{\frac{1}{2}\left[\begin{array}{l}\left(\cancel{y}-\cancel{\frac{1}{2}y^2}+\cancel{\frac{1}{3}y^3}-\cancel{\frac{1}{4}y^4}+\frac{1}{5}y^5+\ldots\right)\\ -\left(-\cancel{y}-\cancel{\frac{1}{2}y^2}-\cancel{\frac{1}{3}y^3}-\cancel{\frac{1}{4}y^4}-\frac{1}{5}y^5+\ldots\right)\end{array}\right]-\cancel{y}-\cancel{\frac{1}{3}y^3}\right\}\\ &=\frac{4m}{5}y^5+\ldots=\frac{4m}{5}\left(\frac{r}{2m}\right)^{5/2}+\ldots=\frac{2}{5}\left(\frac{r}{2m}\right)^{3/2}r+\ldots \end{aligned} \tag{3.218}$$

We notice that in the first small interval of time, $t\simeq 0$, this equation describes an explosion of the central matter, $r\simeq 0$:

$$\frac{r}{t}=c\frac{5}{2}\left(\frac{2m}{r}\right)^{3/2}\xrightarrow{r\to 0}\infty\,, \tag{3.219}$$

i.e. with a matter velocity much larger than the light velocity c, and an inflation of the matter around, with the tendency to empty the black hole. Its matter travels to the Schwarzschild boundary, but reaches this boundary only in an infinite time. According to the velocity and the acceleration (3.206), the matter never passes the Schwarzschild boundary from the inside and also from the outside. Of course, these results, obtained in the framework of a model with spherical symmetry, are violated in the realistic cases when a black hole is submitted to field fluctuations induced by the external and internal bodies, with dynamics determined not only by the gravitational field, but also by much stronger electrical and nuclear phenomena. Due to these fluctuations, the Schwarzschild boundary may be traversed by matter absorption from outside, or by evaporation of matter from inside.

This dynamics suggests that our universe is a huge black hole, which appeared when a large quantity of matter gravitationally collapsed to a spherical structure in the total universe. As we showed above, at that moment, according to the equations (3.217)-(3.219), the central matter undertook a huge explosion, which we call Big Bang. According to equations (3.215) and (3.217), a time evolution began, mainly tending to empty the inner part of the black hole, but with many scattering processes, perturbing this evolution mainly due to the electrical and nuclear interactions. This matter accumulates in the neighborhood of the Schwarzschild boundary of the universe. Supposing a radius $r_0 = 2m = 4.3993 \times 10^{26}$ m of our universe, from (3.163), we obtain the total mass of our universe, $M_G = \frac{c^2}{G} m = 2.9628 \times 10^{53}$ Kg . For a position, let's say $r = m = r_0 / 2$ of a system in universe, we find a universal acceleration (3.206),

$$\frac{\mathrm{d}^2 r}{\mathrm{d}t^2} = -\left[3\left(\frac{2m}{r} - 1\right) + 2\right]\left(\frac{2m}{r} - 1\right)\frac{mc^2}{r^2} = -5\frac{c^2}{m} \approx -5\frac{9\times10^{16}\,\mathrm{m^2 s^{-2}}}{2.2\times10^{26}\,\mathrm{m}} \approx -2\times10^{-9}\,\mathrm{m s^{-2}},$$

which means that such a system far from other celestial bodies is a quasi-inertial one. Since, according to (3.132), the light emitted in a stronger gravitational field is red-shifted, the light emitted in regions nearer the Schwarzschild boundary of our universe, where the matter concentrates, is a red-shifted.

CONCLUSION

We derived fundamental results of the general theory of relativity, as the fundamental equation, the Christoffel symbols, the covariant derivative, the Schwarzschild solution, and Einstein's law of gravitation. We conceived our universe in the four-dimensional time-space physical system as a hypersurface curved in the total universe, including a system of extra-dimensions, which are reduced to the metric tensor. For a gravitational field with spherical symmetry, we integrated the geodesic equations for the radial motion. For a black hole, we considered the external part as a time-like region, where particles move with velocities smaller than the light velocity, and the internal part, where, at the formation of the black hole, the central part of the matter explodes with a velocity much larger than the light velocity. Approaching the Schwarzschild boundary, the matter is decelerated, reaching this boundary only in an infinite time. A quantum particle in the center of a black hole is crushed up in the momentum space, being spread in the coordinate space. In the proximity of the Schwarzschild boundary, a quantum particle is crushed up in the coordinate space, being spread in the momentum space, however, with a null velocity and acceleration. From this perspective, our universe appears as a huge black hole, formed by the accumulation of matter in a structure with spherical symmetry, in the total universe. According to the general theory of relativity, the matter in the center of this black hole exploded with a velocity much larger than the light velocity, but being decelerated and reaching the Schwarzschild boundary only in an infinite time – Big Bang + inflation. The light emitted nearer the Schwarzschild boundary is gravitationally redshifted. For the huge radius of our universe, the acceleration of a body in this universe is much smaller than the acceleration in the gravitational field of Earth – any referential far from any celestial body is a quasi-inertial one.

CHAPTER 4

Charged Quantum Particle in Gravitation and Electromagnetic Fields

Abstract: We describe the interaction of a quantum particle with an electromagnetic field by additional Lagrangian terms in the time dependent phases of the two conjugate wave functions, proportional to the electric charge, with a scalar potential conjugated to time, and a vector potential conjugated to the space coordinates. From the group velocity in the momentum space, we obtain the Lorentz force, as a function of the electric and magnetic fields satisfying the Faraday-Maxwell equation of the electromagnetic induction, and the two Gauss-Maxwell equations for the electric and magnetic fields. When the electromagnetic field is considered as a wave propagating with the maximum relativistic velocity c, the Ampère-Maxwell law and the charge conservation equation are obtained. We consider a gravitational wave for the metric tensor satisfying the D'Alembert equation. For the amplitude tensor of a first-order solution of this equation, we obtain a scalar which describes a rotation of this tensor in a plane perpendicular to the propagation direction of the gravitational wave, with an angular momentum 2, which we call the graviton spin. For the polarization vector we also obtain an invariant describing a matter rotation in this plane, with an integer spin for Bosons, and a half-integer spin for Fermions. The first-order solution describes a particle acceleration in the direction of propagation of the gravitational wave, as the second-order solution describes a harmonic oscillation in this wave. We consider the two propagation wave functions as products of propagation factors, depending only on coordinates and momentum, with time-dependent wave functions satisfying Schrödinger-like time-dependent equations. For a time-dependent wave four-vector, we obtain Dirac-like equations including additional terms explicitly depending on velocity, as it is expected for any relativistic equations. For an electromagnetic decay of a quantum particle, we obtain a redshift depending on the gravitational field. For a free quantum particle with a generalized momentum including the time-dependent component, the time-dependent equations take the form of the quantum field equations with matrix coefficients satisfying the Clifford algebra. We obtain the solutions of these equations for particles and anti-particles, as integrals over the spatial momentum domain, which determines a finite distribution of matter in the coordinate space. Since any matter velocity is equal to the wave velocity, $\dot{\vec{r}} = \frac{\partial}{\partial \vec{p}} cE$, this distribution is invariant.

Keywords: Electric charge, Electric potential, Vector potential, Electric field, Magnetic field, Lorentz force, Maxwell equations, Faraday-Maxwell equation, Ampère-Maxwell equation, Gauss-Maxwell equations, Wave equation, Light

Eliade Stefanescu

velocity, Decay rate, Electric Permittivity, Magnetic permeability, Charge density, Electric current density, Electric induction, Magnetic induction, Particle mobility, Electric conductivity, Conservation equation, D'Alembert equation, Metric tensor, Metric tensor amplitude, Polarization vector, Polarization tensor, Graviton spin, Particle spin, Boson, Fermion, Wave function, Wave four-vector, Redshift, Pauli spin matrix, Dirac spin matrix, Momentum four-vector Clifford algebra, Antiparticle, Covariant normalization.

4.1. QUANTUM PARTICLE DYNAMICS IN ELECTROMAGNETIC FIELD

The interaction with an electromagnetic field, of a charged quantum particle with the wave functions

$$\begin{aligned}\psi\left(x^i,t\right)&=\frac{1}{\left(2\pi\hbar\right)^{3/2}}\int\varphi\left(P^j,t\right)e^{\frac{i}{\hbar}\left[P^jx^j-L\left(x^\alpha,\dot{x}^\alpha\right)t\right]}\mathrm{d}^3P\\ \varphi\left(P^j,t\right)&=\frac{1}{\left(2\pi\hbar\right)^{3/2}}\int\psi\left(x^i,t\right)e^{-\frac{i}{\hbar}\left[P^jx^j-L\left(x^\alpha,\dot{x}^\alpha\right)t\right]}\mathrm{d}^3x,\end{aligned}\tag{4.1}$$

is described by the relativistic Lagrangian with additional term, conjugated to time and to coordinates. For simplicity, we consider a flat space, with the time $t=x^0/c$, and the coordinates $\vec{r}=\left(x^j\right)=\left(x,y,z\right)$. With the potential $U\left(\vec{r}\right)$ conjugated to time, called electric potential, and the potential $\vec{A}\left(\vec{r},t\right)$, conjugated to coordinates, called vector potential, this Lagrangian is

$$L\left(\vec{r},\dot{\vec{r}},t\right)=-\mathrm{M}c^2\sqrt{1-\frac{\dot{r}^2}{c^2}}-eU\left(\vec{r}\right)+e\vec{A}\left(\vec{r},t\right)\dot{\vec{r}}\ ,\tag{4.2}$$

where we consider a time-independent electric potential as of an atom, and a time-dependent vector potential as of and electromagnetic radiation. From this Lagrangian, we obtain the canonical momentum

$$\vec{P}=\frac{\partial}{\partial\dot{\vec{r}}}L\left(\vec{r},\dot{\vec{r}},t\right)=\frac{M\dot{\vec{r}}}{\sqrt{1-\frac{\dot{r}^2}{c^2}}}+e\vec{A}\left(\vec{r},t\right)=\vec{p}+e\vec{A}\left(\vec{r},t\right),\tag{4.3}$$

which besides the mechanical momentum

$$\vec{p}=\frac{M\dot{\vec{r}}}{\sqrt{1-\frac{\dot{\vec{r}}^2}{c^2}}}, \tag{4.4}$$

includes the electromagnetic momentum $e\vec{A}(\vec{r},t)$. With this momentum, we obtain the Hamiltonian

$$\begin{aligned} H(\vec{P},\vec{r},t) &= \vec{P}\dot{\vec{r}}-L(\vec{r},\dot{\vec{r}},t) \\ &= \left[\frac{M\dot{\vec{r}}}{\sqrt{1-\frac{\dot{\vec{r}}^2}{c^2}}}+\underline{e\vec{A}(\vec{r},t)}\right]\dot{\vec{r}}-\left[-\mathrm{M}c^2\sqrt{1-\frac{\dot{\vec{r}}^2}{c^2}}-eU(\vec{r})+\underline{e\vec{A}(\vec{r},t)\dot{\vec{r}}}\right] \\ &= \frac{Mc^2}{\sqrt{1-\frac{\dot{\vec{r}}^2}{c^2}}}+eU(\vec{r})=E(\vec{r},\dot{\vec{r}}), \end{aligned} \tag{4.5}$$

as a conservative function, equal to a time-independent function called energy. With the expression (4.4), for the first term of this expression we obtain the identity

$$\frac{M^2c^2}{1-\frac{\dot{\vec{r}}^2}{c^2}}=\frac{M^2\dot{\vec{r}}^2}{1-\frac{\dot{\vec{r}}^2}{c^2}}+M^2c^2=\vec{p}^2+M^2c^2 .$$

With this expression and the canonical momentum (4.3), the Hamiltonian (4.5) takes the canonical form, as a function of coordinates and momentum:

$$H(\vec{P},\vec{r},t)=c\sqrt{M^2c^2+\vec{p}^2}+eU(\vec{r})=c\sqrt{M^2c^2+\left[\vec{P}-e\vec{A}(\vec{r},t)\right]^2}+eU(\vec{r}). \tag{4.6}$$

At the same time, from (4.3), we obtain electromagnetic force:

$$\vec{F}_e=\frac{\mathrm{d}}{\mathrm{d}t}\vec{p}=\frac{\mathrm{d}}{\mathrm{d}t}\vec{P}-e\frac{\mathrm{d}}{\mathrm{d}t}\vec{A}(\vec{r},t)=\frac{\mathrm{d}}{\mathrm{d}t}\vec{P}-e\frac{\partial}{\partial t}\vec{A}(\vec{r},t)-e\left(\dot{\vec{r}}\frac{\partial}{\partial\vec{r}}\right)\vec{A}(\vec{r},t). \tag{4.7}$$

In this equation, as the derivative of the canonical momentum, we consider the group velocity of the wave-function (4.1) in the momentum space

$$\frac{\mathrm{d}}{\mathrm{d}t}\vec{P}=\frac{\partial}{\partial\vec{r}}L\left(\vec{r},\dot{\vec{r}},t\right)=-e\frac{\partial}{\partial\vec{r}}U\left(\vec{r}\right)+e\frac{\partial}{\partial\vec{r}}\left[\vec{A}\left(\vec{r},t\right)\dot{\vec{r}}\right],$$

where we consider the last term from the vector formula

$$\dot{\vec{r}}\times\left[\frac{\partial}{\partial\vec{r}}\times\vec{A}\left(\vec{r},t\right)\right]=\frac{\partial}{\partial\vec{r}}\left[\dot{\vec{r}}\vec{A}\left(\vec{r},t\right)\right]-\left(\dot{\vec{r}}\frac{\partial}{\partial\vec{r}}\right)\vec{A}\left(\vec{r},t\right).$$

We obtain

$$\frac{\mathrm{d}}{\mathrm{d}t}\vec{P}=\frac{\partial}{\partial\vec{r}}L\left(\vec{r},\dot{\vec{r}},t\right)=-e\frac{\partial}{\partial\vec{r}}U\left(\vec{r}\right)+e\dot{\vec{r}}\times\left[\frac{\partial}{\partial\vec{r}}\times\vec{A}\left(\vec{r},t\right)\right]+\left(\dot{\vec{r}}\frac{\partial}{\partial\vec{r}}\right)\vec{A}\left(\vec{r},t\right).$$

as equation (4.7) takes the form

$$\vec{F}_e=-e\frac{\partial}{\partial\vec{r}}U\left(\vec{r}\right)-e\frac{\partial}{\partial t}\vec{A}\left(\vec{r},t\right)+e\dot{\vec{r}}\times\left[\frac{\partial}{\partial\vec{r}}\times\vec{A}\left(\vec{r},t\right)\right].$$

With the electric field

$$\vec{E}\left(\vec{r},t\right)=-\frac{\partial}{\partial\vec{r}}U\left(\vec{r}\right)-\frac{\partial}{\partial t}\vec{A}\left(\vec{r},t\right), \qquad \textbf{(4.8)}$$

and the magnetic induction

$$\vec{B}\left(\vec{r},t\right)=\frac{\partial}{\partial\vec{r}}\times\vec{A}\left(\vec{r},t\right), \qquad \textbf{(4.9)}$$

as functions of the two potentials $U\left(\vec{r}\right)$ and $\vec{A}\left(\vec{r},t\right)$, we obtain the well-known Lorentz's force

$$\vec{F}_e=e\vec{E}\left(\vec{r},t\right)+e\dot{\vec{r}}\times\vec{B}\left(\vec{r},t\right). \qquad \textbf{(4.10)}$$

Since the curl of the gradient is null, from (4.8) with (4.9), we obtain the Faraday-Maxwell law of the electromagnetic induction:

$$\frac{\partial}{\partial \vec{r}} \times \vec{E}(\vec{r},t) = -\frac{\partial}{\partial t}\vec{B}(\vec{r},t). \qquad \textbf{(4.11)}$$

Since the divergence of the curl is null, from (4.9) we obtain the Gauss-Maxwell law of the magnetic induction:

$$\frac{\partial}{\partial \vec{r}}\vec{B}(\vec{r},t) = 0. \qquad \textbf{(4.12)}$$

With the gouge condition for the vector potential,

$$\frac{\partial}{\partial \vec{r}}\vec{A}(\vec{r},t) = 0, \qquad \textbf{(4.13)}$$

from (4.8), we obtain the Gauss-Maxwell law of the electric field,

$$\frac{\partial}{\partial \vec{r}}\vec{E}(\vec{r},t) = -\frac{\partial^2}{\partial \vec{r}^2}U(\vec{r},t) = \frac{\rho(\vec{r},t)}{\varepsilon_0}, \qquad \textbf{(4.14)}$$

with the electric charge density

$$\rho(\vec{r},t) = \varepsilon_0 \frac{\partial}{\partial \vec{r}}\vec{E}(\vec{r},t) = -\varepsilon_0 \frac{\partial^2}{\partial \vec{r}^2}U(\vec{r},t) \qquad \textbf{(4.15)}$$

as a source of this field, with the coefficient ε_0 as a dimensional constant called electric permittivity. From (4.12) with the Gauss theorem, we obtain a total null flux of the magnetic induction through a closed surface Σ :

$$\oint_{\Sigma} \vec{B}(\vec{r},t)\mathrm{d}^2\vec{r} = 0. \qquad \textbf{(4.16)}$$

From (4.14) with the Gauss theorem, we obtain the total flux of the electric field through a closed surface Σ as the total electric charge q in the volume V_Σ with this surface:

$$\oint_\Sigma \vec{E}(\vec{r},t)\mathrm{d}^2\vec{r} = \frac{1}{\varepsilon_0}\int_{V_\Sigma} \rho(\vec{r},t)\mathrm{d}^3\vec{r} = \frac{q}{\varepsilon_0}. \quad \textbf{(4.17)}$$

For a system with spherical symmetry, from this equation, we obtain the Coulomb formula of a central electrical field:

$$\vec{E}(\vec{r},t) = \frac{q}{4\pi\varepsilon_0 r^2}\frac{\vec{r}}{r} = \frac{1}{4\pi\varepsilon_0 r^2}\frac{\vec{r}}{r}\int_{V_\Sigma} \rho(\vec{r},t)\mathrm{d}^3\vec{r}. \quad \textbf{(4.18)}$$

We notice that from equations (4.11), (4.14), and (4.12), we obtain the curl and the divergence of the electric field, and the divergence of the magnetic induction. To obtain the curl of the magnetic induction, we consider the curl of equation (4.11):

$$\frac{\partial}{\partial\vec{r}}\times\left[\frac{\partial}{\partial\vec{r}}\times\vec{E}(\vec{r},t)\right] = -\frac{\partial}{\partial t}\frac{\partial}{\partial\vec{r}}\times\vec{B}(\vec{r},t),$$

which is

$$\frac{\partial}{\partial\vec{r}}\left[\frac{\partial}{\partial\vec{r}}\vec{E}(\vec{r},t)\right] - \frac{\partial^2}{\partial\vec{r}^2}\vec{E}(\vec{r},t) = -\frac{\partial}{\partial t}\frac{\partial}{\partial\vec{r}}\times\vec{B}(\vec{r},t). \quad \textbf{(4.19)}$$

We consider the electric field $\vec{E}(\vec{r},t)$ as a wave propagating with the velocity c and a dissipative decay rate γ in the environment with the charge density $\rho(\vec{r},t)$,

$$\frac{\partial^2}{\partial\vec{r}^2}\vec{E}(\vec{r},t) = \frac{1}{c^2}\left[\frac{\partial^2}{\partial t^2}\vec{E}(\vec{r},t) + \gamma\frac{\partial}{\partial t}\vec{E}(\vec{r},t)\right]. \quad \textbf{(4.20)}$$

From these two equations, we obtain

$$\frac{\partial}{\partial t}\frac{\partial}{\partial \vec{r}}\times\vec{B}(\vec{r},t)=\frac{1}{c^2}\left[\frac{\partial^2}{\partial t^2}\vec{E}(\vec{r},t)+\gamma\frac{\partial}{\partial t}\vec{E}(\vec{r},t)\right]-\frac{\partial}{\partial \vec{r}}\left[\frac{\partial}{\partial \vec{r}}\vec{E}(\vec{r},t)\right].$$

Since a particle with the wave functions (4.1) is very small, we neglect the spatial variation of the charge density, as the last term of this equation is null. We obtain a first-order differential equation in space and time:

$$\frac{\partial}{\partial \vec{r}}\times\vec{B}(\vec{r},t)=\frac{1}{c^2}\left[\frac{\partial}{\partial t}\vec{E}(\vec{r},t)+\gamma\vec{E}(\vec{r},t)\right].$$

This equation with the dimensional constant μ_0, called magnetic permeability, which satisfies the relation

$$c=\frac{1}{\sqrt{\varepsilon_0\mu_0}}, \qquad \textbf{(4.21)}$$

takes the form of the Ampère-Maxwell law,

$$\frac{1}{\mu_0}\frac{\partial}{\partial \vec{r}}\times\vec{B}(\vec{r},t)=\varepsilon_0\gamma\vec{E}(\vec{r},t)+\varepsilon_0\frac{\partial}{\partial t}\vec{E}(\vec{r},t), \qquad \textbf{(4.22)}$$

of the curl of the magnetic field

$$\vec{H}(\vec{r},t)=\frac{1}{\mu_0}\vec{B}(\vec{r},t) \qquad \textbf{(4.23)}$$

as a sum of a term proportional to the electric field, called the electric current density,

$$\vec{j}(\vec{r},t)=\varepsilon_0\gamma\vec{E}(\vec{r},t), \qquad \textbf{(4.24)}$$

and the time derivative of a quantity called the electric induction,

$$\vec{D}(\vec{r},t)=\varepsilon_0\vec{E}(\vec{r},t). \qquad \textbf{(4.25)}$$

With these notations, the Gauss-Maxwell law (4.14) and the Ampère-Maxwell law (4.22) take the simple forms

$$\frac{\partial}{\partial \vec{r}} \vec{D}(\vec{r},t) = \rho(\vec{r},t), \tag{4.26}$$

and

$$\frac{\partial}{\partial \vec{r}} \times \vec{H}(\vec{r},t) = \vec{j}(\vec{r},t) + \frac{\partial}{\partial t} \vec{D}(\vec{r},t). \tag{4.27}$$

For a clearer understanding of the charge current density (4.24), for the right-hand side of equation (4.22), we consider the total derivative of the electrical field:

$$\begin{aligned} &\underline{\varepsilon_0 \frac{\partial}{\partial t} \vec{E}(\vec{r},t)} + \varepsilon_0 \gamma \vec{E}(\vec{r},t) = \varepsilon_0 \left[\frac{\mathrm{d}}{\mathrm{d}t} \vec{E}(\vec{r},t) \right] \\ &= \varepsilon_0 \left[\underline{\frac{\partial}{\partial t} \vec{E}(\vec{r},t)} + \left(\dot{\vec{r}} \frac{\partial}{\partial \vec{r}} \right) \vec{E}(\vec{r},t) \right]. \end{aligned} \tag{4.28}$$

We obtain the last term of this equation from the double vector product identity,

$$\frac{\partial}{\partial \vec{r}} \times \left[\dot{\vec{r}} \times \vec{E}(\vec{r},t) \right] = \dot{\vec{r}} \left[\frac{\partial}{\partial \vec{r}} \vec{E}(\vec{r},t) \right] - \left(\dot{\vec{r}} \frac{\partial}{\partial \vec{r}} \right) \vec{E}(\vec{r},t),$$

considering the particle velocity in the direction of the electromagnetic field: $\dot{\vec{r}} \times \vec{E}(\vec{r},t) = 0$.

With this equation, (4.15), and (4.28), from (4.24) we obtain the electric current density

$$\vec{j}(\vec{r},t) = \dot{\vec{r}} \rho(\vec{r},t), \tag{4.29}$$

as the product of the charge density with the velocity. From this expression with (4.24), we obtain the particle velocity in the electric field:

$$\dot{\vec{r}}\frac{\rho}{\varepsilon_0}=\gamma\vec{E}(\vec{r},t).$$

Considering the statistical equation of the particle velocity with the mobility μ, as a function of the electrical field inducing this velocity [7],

$$\dot{\vec{r}}=\mu\vec{E}(\vec{r},t), \tag{4.30}$$

we obtain the decay rate γ of the field equation (4.20) as a function of charge density ρ and particle mobility μ:

$$\gamma=\frac{\mu}{\varepsilon_0}\rho. \tag{4.31}$$

With the mobility equation (4.30), we obtain the electric current density (4.29),

$$\begin{gathered}\vec{j}(\vec{r},t)=\rho\dot{\vec{r}}=\mu\rho\vec{E}(\vec{r},t)=\sigma\vec{E}(\vec{r},t),\\ \left[\mathrm{C\,m^{-3}\,m\,s^{-1}}\right]=\left[\mathrm{C\,s^{-1}\,m^{-2}}\right]=\left[\mathrm{A\,m^{-2}}\right]\end{gathered}, \tag{4.32}$$

as a function of the electric field and the electric conductivity

$$\sigma=\mu\rho. \tag{4.33}$$

From equation (4.31) with (4.33), we obtain the decay rate of the electromagnetic field as a function of the electric conductivity:

$$\gamma=\frac{\sigma}{\varepsilon_0},\quad\left[\frac{\cancel{\mathrm{C}}\,\mathrm{s}^{-1}\,\cancel{\mathrm{m}}^{-2}/\cancel{\mathrm{V}}\,\cancel{\mathrm{m}}^{-1}}{\cancel{\mathrm{F}}\,\cancel{\mathrm{m}}^{-1}}\right]=\left[\mathrm{s}^{-1}\right]. \tag{4.34}$$

Since the divergence of a curl is null, from (4.26) and (4.27), we obtain the equation of conservation of the electric charge:

$$\frac{\partial}{\partial\vec{r}}\vec{j}(\vec{r},t)=-\frac{\partial}{\partial t}\rho(\vec{r},t). \tag{4.35}$$

In this way, from the interaction of an electromagnetic field with a quantum particle (4.1) with the relativistic Lagrangian (4.2), we obtained Lorentz's force (4.10), the equation of conservation (4.35) of the electric charge as a source of the electric field (4.15), and the Maxwell equations. The Faraday-Maxwell law of electromagnetic induction (4.11) describes the electric field as a curl induced by the time variation of the magnetic induction. The Gauss-Maxwell law (4.12) describes a null divergence of the magnetic induction. From the Gauss-Maxwell law (4.26), we obtain the divergence of the electric induction as the electric charge density. The Ampère-Maxwell law (4.27) describes the curl of the magnetic field as the sum of the electric current density with the time variation of the electric induction. These equations for the magnetic field $\vec{H}(\vec{r},t)$ and the electric field $\vec{E}(\vec{r},t)$ are:

$$\begin{aligned}
&\frac{\partial}{\partial\vec{r}}\times\vec{H}(\vec{r},t)=\vec{j}(\vec{r},t)+\varepsilon_0\frac{\partial}{\partial t}\vec{E}(\vec{r},t)\\
&\frac{\partial}{\partial\vec{r}}\vec{H}(\vec{r},t)=0\\
&\frac{\partial}{\partial\vec{r}}\times\vec{E}(\vec{r},t)=-\mu_0\frac{\partial}{\partial t}\vec{H}(\vec{r},t)\\
&\frac{\partial}{\partial\vec{r}}\vec{E}(\vec{r},t)=\frac{\rho(\vec{r},t)}{\varepsilon_0}.
\end{aligned} \tag{4.36}$$

According to (4.21), we can choose the two-dimensional constants ε_0 , and μ_0 as

$$\varepsilon_0=\mu_0=\sqrt{\varepsilon_0\mu_0}=\frac{1}{c}. \tag{4.37}$$

To obtain a more convenient form for the theory of relativity of the Maxwell equations [2], we use a normalized charge density

$$\tilde{\rho}(\vec{r},t)=\frac{\rho(\vec{r},t)}{4\pi\varepsilon_0}=\frac{c\rho(\vec{r},t)}{4\pi}, \tag{4.38}$$

leading to a simpler form of the Coulomb law (4.18)

$$\vec{E}(\vec{r},t)=\frac{1}{r^2}\frac{\vec{r}}{r}\int_{V_\Sigma}\tilde{\rho}(\vec{r},t)\mathrm{d}^3\vec{r}\,, \tag{4.39}$$

and the normalized current density according to (4.29):

$$\vec{j}(\vec{r},t)=\dot{\vec{r}}\rho(\vec{r},t)=4\pi\frac{\dot{\vec{r}}}{c}\tilde{\rho}(\vec{r},t)=4\pi\tilde{\vec{j}}(\vec{r},t), \tag{4.40}$$

We notice that the normalized current density is the product of the normalized density with the velocity on time relativistic coordinate $x^0=ct$:

$$\tilde{\vec{j}}(\vec{r},t)=\frac{\dot{\vec{r}}}{c}\tilde{\rho}(\vec{r},t)=\frac{1}{c}\frac{\mathrm{d}\vec{r}}{\mathrm{d}t}\tilde{\rho}(\vec{r},t)=\tilde{\rho}(\vec{r},t)\frac{\mathrm{d}\vec{r}}{\mathrm{d}x^0}. \tag{4.41}$$

In this way, the Maxwell equations (4.36) take the form of a system of equations with the same dimensions for the electric and magnetic fields, for the electric permittivity and the magnetic permeability, and the electric charge and electric current densities:

$$\begin{aligned}
&\frac{1}{c}\frac{\partial}{\partial t}\vec{E}(\vec{r},t)=\frac{\partial}{\partial\vec{r}}\times\vec{H}(\vec{r},t)-4\pi\tilde{\vec{j}}(\vec{r},t)\\
&\frac{\partial}{\partial\vec{r}}\vec{E}(\vec{r},t)=4\pi\tilde{\rho}(\vec{r},t)\\
&\frac{1}{c}\frac{\partial}{\partial t}\vec{H}(\vec{r},t)=-\frac{\partial}{\partial\vec{r}}\times\vec{E}(\vec{r},t)\\
&\frac{\partial}{\partial\vec{r}}\vec{H}(\vec{r},t)=0.
\end{aligned} \tag{4.42}$$

4.2. QUANTUM PARTICLE IN A GRAVITATIONAL WAVE AND THE GRAVITON SPIN

For the particle dynamics in a gravitational wave, generated only by far heavy celestial bodies, we consider low velocities

$$\dot{x}^i<<\dot{x}^0\simeq 1, \tag{4.43}$$

for the wave functions

$$\psi\left(x^i,t\right)=\frac{1}{\left(2\pi\hbar\right)^{3/2}}\int\varphi\left(P^j,t\right)e^{\frac{i}{\hbar}\left[P^jx^j-L\left(x^\alpha,v^\alpha\right)t\right]}\mathrm{d}^3P$$
$$\varphi\left(P^j,t\right)=\frac{1}{\left(2\pi\hbar\right)^{3/2}}\int\psi\left(x^i,t\right)e^{-\frac{i}{\hbar}\left[P^jx^j-L\left(x^\alpha,v^\alpha\right)t\right]}\mathrm{d}^3x, \tag{4.44}$$

with the Lagrangian

$$L\left(x^\alpha,\dot{x}^\alpha\right)=-Mc^2\sqrt{g_{\alpha\beta}\dot{x}^\alpha\dot{x}^\beta}, \tag{4.45}$$

and the momentum

$$P^j=\frac{\partial L}{c\partial\dot{x}^j}=-Mcg_{ij}\dot{x}^i. \tag{4.46}$$

We notice that with the inequality (4.43), the fundamental equation (1.48) takes the form

$$g_{\alpha\beta}\dot{x}^\alpha\dot{x}^\beta=1. \tag{4.47}$$

With this relation, the group velocity of the wave-packet (4.44) with the Lagrangian (4.45) in the momentum space is:

$$\frac{\mathrm{d}}{\mathrm{d}t}P^j=c\dot{P}^j=-Mc^2g_{ij,k}\dot{x}^k\dot{x}^i-Mc^2g_{ij}\ddot{x}^i=\frac{\partial L}{\partial x^j}$$
$$=-Mc^2\frac{\dot{x}^\alpha\dot{x}^\beta g_{\alpha\beta,j}}{2\sqrt{g_{\alpha\beta}\dot{x}^\alpha\dot{x}^\beta}}=-\frac{1}{2}Mc^2g_{\alpha\beta,j}\dot{x}^\alpha\dot{x}^\beta.$$

For the case (4.43) of the low velocities, from this expression multiplied with g^{kj}, we obtain the acceleration vector as a gradient of the metric element g_{00}:

$$\ddot{x}^k = \frac{1}{2} g^{kj} g_{00,j} \,. \quad \textbf{(4.48)}$$

We notice that these coordinate oscillations, which satisfy the fundamental condition (4.47), are correlated with oscillations of the metric elements in the gravitational wave according to the D'Alembert equation

$$g^{\mu\nu} g_{\rho\sigma,\mu\nu} = 0 \,. \quad \textbf{(4.49)}$$

We consider a solution of this equation in the first-order approximation

$$g_{\rho\sigma} = u_{\rho\sigma} l_\mu x^\mu \,, \quad \textbf{(4.50)}$$

with the symmetric amplitude tensor

$$u_{\rho\sigma} = u g_{\rho\sigma} \,, \quad u^{\rho\sigma} = u g^{\rho\sigma} \quad \textbf{(4.51)}$$

as a function of the scalar amplitude u, and the polarization vector l_μ, which satisfies the normalization condition

$$l^\nu l_\nu = g^{\mu\nu} l_\mu l_\nu = 0 \,. \quad \textbf{(4.52)}$$

With the expression (4.50) of the metric tensor, for the particle acceleration (4.48), we find a constant value in the direction l^k of the propagation of the gravitational wave:

$$\ddot{x}^k = \frac{1}{2} g^{kj} u_{00} l_j = \frac{1}{2} u_{00} l^k \,. \quad \textbf{(4.53)}$$

By multiplying the second equation (4.51) with $g_{\mu\rho}$,

$$g_{\mu\rho} u^{\rho\sigma} = u g_{\mu\rho} g^{\rho\sigma} \,,$$

we obtain the amplitude tensor as a function only of the scalar amplitude

$$u_{\mu}^{\sigma} = u\delta_{\mu}^{\sigma} . \tag{4.54}$$

By multiplying this expression with the polarization vector, we obtain a system of equations for the elements of the amplitude tensor:

$$l_{\sigma} u_{\mu}^{\sigma} = l_{\mu} u . \tag{4.55}$$

By contracting the indices of this tensor, from (4.54) we find an equation for its diagonal elements:

$$u_{\mu}^{\mu} = 4u . \tag{4.56}$$

For the realistic case of a weak gravitational wave

$$\begin{aligned} g_{00} = 1, \quad g_{11} = -1, \quad g_{22} = -1, \quad g_{33} = -1 \\ g^{00} = 1, \quad g^{11} = -1, \quad g^{22} = -1, \quad g^{33} = -1, \end{aligned} \tag{4.57}$$

with the amplitude element (4.51)band the fundamental equation (4.47),

$$u_{00} = u g_{00} = u \tag{4.58}$$

for the dynamic equation (4.53) we obtain a form depending on the scalar amplitude u,

$$\ddot{x}^{k} = \frac{1}{2} u l^{k} . \tag{4.59}$$

We consider a gravitational wave propagating in the direction x^{3}, with the polarization vector

$$\begin{aligned} l_{0} = 1, \quad l_{1} = l_{2} = 0, \quad l_{3} = -1 \\ l^{0} = 1, \quad l^{1} = l^{2} = 0, \quad l^{3} = 1, \end{aligned} \tag{4.60}$$

which satisfies the normalization relation (4.52). For tis polarization vector, equations (4.55) take the explicit form:

$$\begin{aligned}
u_0^0 - u_0^3 &= g^{00}u_{00} - g^{33}u_{30} = u_{00} + u_{30} = u \\
u_1^0 - u_1^3 &= g^{00}u_{01} - g^{33}u_{31} = u_{01} + u_{31} = 0 \\
u_2^0 - u_2^3 &= g^{00}u_{02} - g^{33}u_{32} = u_{02} + u_{32} = 0 \\
u_3^0 - u_3^3 &= g^{00}u_{03} - g^{33}u_{33} = u_{03} + u_{33} = -u .
\end{aligned} \quad \textbf{(4.61)}$$

With the symmetry property of the amplitude tensor, and (4.51) with (4.57), we obtain

$$\begin{aligned}
u_{00} + u_{33} &= -2u_{03} \\
u_{00} - u_{33} &= 2u = -\left(u_{11} + u_{22}\right) \\
u_{10}{}^2 &= u_{13}{}^2 \\
u_{20}{}^2 &= u_{23}{}^2 .
\end{aligned} \quad \textbf{(4.62)}$$

We define the scalar invariant

$$I_0 = u_{\alpha\beta}u^{\alpha\beta} - 4u^2 . \quad \textbf{(4.63)}$$

With the expression

$$u^{\alpha\beta} = g^{\alpha\mu}g^{\beta\nu}u_{\mu\nu} = g^{\alpha\alpha}g^{\beta\beta}u_{\alpha\beta} ,$$

and the relations (4.62), we obtain

$$
\begin{aligned}
I_0 &= u_{\alpha\beta}u^{\alpha\beta} - 2u^2 - 2u^2 \\
&= u_{00}{}^2 + \underline{u_{11}{}^2 + u_{22}{}^2} + u_{33}{}^2 - \underline{\underline{\cancel{2u_{10}{}^2}}} - \underline{\underline{\cancel{2u_{20}{}^2}}} - 2u_{03}{}^2 + \underline{2u_{12}{}^2} + \underline{\underline{\cancel{2u_{23}{}^2}}} + \underline{\underline{\cancel{2u_{13}{}^2}}} \\
&\quad - \frac{1}{2}\left(u_{00} - u_{33}\right)^2 - 2u^2 \\
&= u_{11}{}^2 + u_{22}{}^2 + 2u_{12}{}^2 + \underline{u_{00}{}^2} + \underline{u_{33}{}^2} - \underline{\frac{1}{2}\left(u_{00} + u_{33}\right)^2} - \underline{\frac{1}{2}\left(u_{00} - u_{33}\right)^2} - 2u^2 \\
&= u_{11}{}^2 + u_{22}{}^2 + 2u_{12}{}^2 - \frac{1}{2}\left(u_{11} + u_{22}\right)^2 \\
&= \frac{1}{2}\left(u_{11} - u_{22}\right)^2 + 2u_{12}{}^2.
\end{aligned}
\tag{4.64}
$$

This invariant describes a rotation in the plane $\left(x^1, x^2\right)$, perpendicular to the propagation direction x^3. In this plane we consider a vector $A = \left(A^1, A^2\right)$, and a rotation of this plane with an angle $\pi/2$, which means a rotation of this vector with an angle $-\pi/2$, $R_{-\pi/2}A = \left(R_{-\pi/2}A^1, R_{-\pi/2}A^2\right) = \left(A^2, -A^1\right)$.

With a new rotation with an angle $-\pi/2$, we obtain

$$
R_{-\pi}A = R_{-\pi/2}{}^2 A = R_{-\pi/2}\left(A^2, -A^1\right) = \left(-A^1, -A^2\right) = -A,
$$

which means the eigenvalues $R_{-\pi} = -1$ and $R_{-\pi/2} = \pm \mathrm{i}$. With a rotation $R_{-\pi/2}$ of the amplitude tensor, we obtain

$$
\begin{aligned}
R_{-\pi/2}u_{11} &= u_{12} = u_{21} = \frac{1}{2}\left(u_{12} + u_{21}\right) \\
R_{-\pi/2}u_{22} &= -u_{21} = -u_{12} = -\frac{1}{2}\left(u_{12} + u_{21}\right) \\
R_{-\pi/2}u_{21} &= R_{-\pi/2}u_{12} = u_{22} = -u_{11} = \frac{1}{2}\left(u_{22} - u_{11}\right).
\end{aligned}
\tag{4.65}
$$

From the first two equations we obtain

$$R_{-\pi/2}(u_{11}+u_{22})=0$$
$$R_{-\pi/2}(u_{11}-u_{22})=2u_{12}. \quad \mathbf{(4.66)}$$

From the second equation (4.66) with the third equation (4.65), we obtain the rotation property

$$R_{-\pi}(u_{11}-u_{22})=R_{-\pi/2}{}^{2}(u_{11}-u_{22})=R_{-\pi/2}(2u_{12})$$
$$=u_{22}-u_{11}, \quad \mathbf{(4.67)}$$

which means that for a rotation with the angle $-\pi$ the first term of the invariant (4.64) retakes its value. From the third equation (4.65) with the second equation (4.66), we obtain

$$R_{-\pi}(2u_{12})=R_{-\pi/2}{}^{2}(2u_{12})=R_{-\pi/2}(u_{22}-u_{11})$$
$$=-2u_{12}, \quad \mathbf{(4.68)}$$

which means that for a rotation with the angle $-\pi$, also the second term of the invariant (4.64) retakes its value. From (4.65) and (4.66), we notice that with a rotation of the angle $-\pi/2$, the two terms of the invariant (4.64) transform one another:

$$\frac{1}{2}\left[R_{-\pi/2}(u_{11}-u_{22})\right]^{2}=\frac{1}{2}(2u_{12})^{2}=2u_{12}{}^{2}$$
$$2(R_{-\pi/2}u_{12})^{2}=2\left[\frac{1}{2}(u_{22}-u_{11})\right]^{2}=\frac{1}{2}(u_{11}-u_{22})^{2}. \quad \mathbf{(4.69)}$$

We consider a rotation $R_{\delta\vec{\alpha}}=e^{\mathrm{i}\vec{S}\delta\vec{\alpha}}$ with an arbitrary angle $\delta\vec{\alpha}$, as a function of the angular momentum

$$\vec{S}=-\mathrm{i}\vec{r}\times\frac{\partial}{\partial\vec{r}}. \quad \mathbf{(4.70)}$$

We notice that a rotation with an angle $-\pi$ with an eigenvalue -1,

$$R_{-\pi}A(\vec{r}) = e^{-\mathrm{i}S\pi}A(\vec{r}) = -A(\vec{r})$$

is obtained for an angular momentum $S = 1$, which we call spin. To obtain the spin S_t of the tensor u_{ij}, we consider the scalar product of this tensor with two vectors, which is invariant:

$$\begin{aligned}\text{Scalar} = u_{ij}A^iB^j &= \left[R^{\delta\vec{\alpha}}u_{ij}\right]\left[R^{\delta\vec{\alpha}}A^i\right]\left[R^{\delta\vec{\alpha}}B^j\right] \\ &= \left[e^{\mathrm{i}\vec{S}_t\delta\vec{\alpha}}u_{ij}\right]\left[e^{\mathrm{i}\vec{S}\delta\vec{\alpha}}A^i\right]\left[e^{\mathrm{i}\vec{S}\delta\vec{\alpha}}B^j\right] \\ &= \left[e^{\mathrm{i}\vec{S}_t\delta\vec{\alpha}}u_{ij}\right]\left[e^{\mathrm{i}\delta\alpha}A^i\right]\left[e^{\mathrm{i}\delta\alpha}B^j\right].\end{aligned}$$

We obtain the tensor spin eigenvalue

$$S_t = 2, \tag{4.71}$$

which we call the graviton spin.

We notice that the time-space invariant (4.47) with the metric tensor (4.50) is the product of two invariants,

$$g_{\alpha\beta}\dot{x}^\alpha\dot{x}^\beta = u_{\alpha\beta}l_\mu x^\mu\dot{x}^\alpha\dot{x}^\beta = I_G I_C = 1, \tag{4.72}$$

a graviton invariant of the metric tensor, which with (4.51) is the scalar amplitude of the metric tensor,

$$I_G = u_{\alpha\beta}\dot{x}^\alpha\dot{x}^\beta = ug_{\alpha\beta}\dot{x}^\alpha\dot{x}^\beta = u, \tag{4.73}$$

and the coordinate invariant, which is the inverse of this amplitude,

$$I_C = l_\mu x^\mu = \frac{1}{u}. \tag{4.74}$$

With the invariant (4.64), the graviton invariant (4.73) describes a rotation with the spin 2 in the plane $\left(x^1, x^2\right)$ for a propagation direction x^3, of the amplitude tensor

$u_{\alpha\beta}$ correlated to the velocity product $\dot{x}^{\alpha}\dot{x}^{\beta}$. With the normalization invariant (4.52), $l_1^2 + l_2^2 = l_0^2 - l_3^2$, the coordinate invariant (4.74) describes a rotation with a spin S in the plane $\left(x^1, x^2\right)$ perpendicular to the propagation direction x^3. For a rotation with the angle 2π, the two terms retake their values,

$$l_1^2 + l_2^2 = \left(R_{2\pi} l_1\right)^2 + (R_{2\pi} l_2)^2 = l_0^2 - l_3^2, \tag{4.75}$$

with the eigenvalues $R_{2\pi} = e^{i2\pi S} = \pm 1$, which means a spin $S = 1$ for Bosons and $S = 1/2$ for Fermions.

For the wave equation (4.49), we can also consider a second-order solution,

$$g_{\rho\sigma} = u_{\rho\sigma} l_{\mu\nu} x^{\mu} x^{\nu}, \tag{4.76}$$

with a polarization tensor $l_{\mu\nu}$. With this solution, from the wave equation (4.49) we obtain

$$g^{\mu\nu} u_{\rho\sigma} l_{\mu\nu} = u_{\rho\sigma} l_{\nu}^{\nu} = 0$$

which leads to the normalization condition for the polarization tensor,

$$l_{\nu}^{\nu} = 0. \tag{4.77}$$

With this solution, the dynamic equation (4.48) becomes

$$\ddot{x}^k = \frac{1}{2} g^{kj} g_{00,j} = \frac{1}{2} g^{kj} u_{00} l_{\mu j} x^{\mu} = \frac{1}{2} u_{00} l_{\mu}^{k} x^{\mu}. \tag{4.78}$$

For a gravitational field oscillating in the direction x^1, which satisfies the normalization condition (4.77),

$$l_0^0 = 1, \quad l_1^1 = -1, \quad l_2^2 = l_3^3 = 0, \tag{4.79}$$

and the expression (4.58) of the scalar amplitude, we obtain a particle harmonic oscillation in the direction of the gravitational field oscillation:

$$\ddot{x}^1 = -\frac{1}{2}ux^1. \tag{4.80}$$

Thus, the first-order solution (4.50) of the gravitational wave equation (4.49) describes the propagation of a quantum particle in this wave according to the dynamic equation (4.59), as the second-order solution (4.76) describes an oscillation in this wave according to equation (4.80). The gravitational field is characterized by a rotation of the metric tensor, coupled with the velocity field of the quantum matter, with the spin 2. The quantum matter dynamics is characterized by a rotation with an integer spin for Bosons, or a half-integer spin for Fermions.

4.3. PROPAGATION WAVE FUNCTIONS

We consider the wave functions (4.1) under the form

$$\begin{aligned}
\psi(\vec{r},t) &= \frac{1}{(2\pi\hbar)^{3/2}}\int \varphi(\vec{p},t)e^{\frac{i}{\hbar}\left[\vec{p}\vec{r}-L(\vec{r},\dot{\vec{r}})t\right]}\mathrm{d}^3\vec{p} = \frac{1}{(2\pi\hbar)^{3/2}}\int \varphi(\vec{p},t)e^{\frac{i}{\hbar}\left\{\vec{p}\vec{r}-\left[\vec{p}\dot{\vec{r}}-H(\vec{p},\vec{r})\right]t\right\}}\mathrm{d}^3\vec{p} \\
&= e^{\frac{i}{\hbar}\hat{\vec{p}}\vec{r}}\frac{1}{(2\pi\hbar)^{3/2}}\int \varphi(\vec{p},t)e^{-\frac{i}{\hbar}\left[\vec{p}\dot{\vec{r}}-H(\vec{p},\vec{r})\right]t}\mathrm{d}^3\vec{p} \\
&= e^{\frac{i}{\hbar}\hat{\vec{p}}\vec{r}}\psi_t(\vec{r},t) \\
\varphi(\vec{p},t) &= \frac{1}{(2\pi\hbar)^{3/2}}\int \psi(\vec{r},t)e^{-\frac{i}{\hbar}\left[\vec{p}\vec{r}-L(\vec{r},\dot{\vec{r}})t\right]}\mathrm{d}^3\vec{r} = \frac{1}{(2\pi\hbar)^{3/2}}\int \psi(\vec{r},t)e^{-\frac{i}{\hbar}\left\{\vec{p}\vec{r}-\left[\vec{p}\dot{\vec{r}}-H(\vec{p},\vec{r})\right]t\right\}}\mathrm{d}^3\vec{r} \\
&= e^{-\frac{i}{\hbar}\vec{p}\hat{\vec{r}}}\frac{1}{(2\pi\hbar)^{3/2}}\int \psi(\vec{r},t)e^{\frac{i}{\hbar}\left[\vec{p}\dot{\vec{r}}-H(\vec{p},\vec{r})\right]t}\mathrm{d}^3\vec{r} \\
&= e^{-\frac{i}{\hbar}\vec{p}\hat{\vec{r}}}\varphi_t(\vec{p},t),
\end{aligned} \tag{4.81}$$

with the propagation factors $e^{\frac{i}{\hbar}\hat{\vec{p}}\vec{r}}$, as an operator in the coordinate space, and $e^{-\frac{i}{\hbar}\vec{p}\hat{\vec{r}}}$, as an operator in the momentum space, of the time-dependent wave functions $\psi_t(\vec{r},t)$ in the coordinate space, and $\varphi_t(\vec{p},t)$ the momentum space. From the explicit forms of the time-dependent wave functions, we obtain the Schrödinger-like Hamiltonian equations:

$$\begin{aligned} &\mathrm{i}\hbar\frac{\partial}{\partial t}\psi_t(\vec{r},t)=\left[\vec{p}\dot{\vec{r}}-H(\vec{p},\vec{r})\right]\psi_t(\vec{r},t)\\ &\mathrm{i}\hbar\frac{\partial}{\partial t}\varphi_t(\vec{p},t)=-\left[\vec{p}\dot{\vec{r}}-H(\vec{p},\vec{r})\right]\varphi_t(\vec{p},t). \end{aligned} \tag{4.82}$$

We consider the Hamiltonian (4.6) under the linear Dirac's form

$$\begin{aligned} H(\vec{P},\vec{r},t)&=c\sqrt{M^2c^2+\vec{p}^2}+eU(\vec{r})\\ &=c(\alpha_0 Mc+\alpha_1 p_1+\alpha_2 p_2+\alpha_3 p_3)+eU(\vec{r}), \end{aligned} \tag{4.83}$$

depending on Dirac spin operators α_μ, $\mu=0,1,2,3$,

$$\alpha_0=\begin{pmatrix}\hat{1} & 0\\ 0 & -\hat{1}\end{pmatrix},\ \alpha_1=\begin{pmatrix}0 & \sigma_1\\ \sigma_1 & 0\end{pmatrix},\ \alpha_2=\begin{pmatrix}0 & \sigma_2\\ \sigma_2 & 0\end{pmatrix},\ \alpha_3=\begin{pmatrix}0 & \sigma_3\\ \sigma_3 & 0\end{pmatrix}, \tag{4.84}$$

as a functions of the Pauli spin operators

$$\sigma_1=\begin{pmatrix}0 & 1\\ 1 & 0\end{pmatrix},\ \sigma_2=\begin{pmatrix}0 & -\mathrm{i}\\ \mathrm{i} & 0\end{pmatrix},\ \sigma_3=\begin{pmatrix}1 & 0\\ 0 & -1\end{pmatrix}, \tag{4.85}$$

which satisfy the anticommutation relations:

$$\{\alpha_\mu,\alpha_\nu\}=2\delta_{\mu\nu},\qquad \{\sigma_i,\sigma_j\}=2\delta_{ij},\qquad \sigma_i\sigma_j=\mathrm{i}\delta_{ijk}\sigma_k. \tag{4.86}$$

In this framework, a quantum particle is described by a wave four-vector

$$\psi_t=\begin{pmatrix}\psi_1\\ \psi_2\end{pmatrix},\qquad \psi_1=\begin{pmatrix}\varphi_1\\ \varphi_2\end{pmatrix},\qquad \psi_2=\begin{pmatrix}\varphi_3\\ \varphi_4\end{pmatrix}. \tag{4.87}$$

From the first wave equation (4.82) we obtain a system of equations like Dirac's equations, but with additional terms depending on velocity:

$$-\mathrm{i}\hbar\left(\frac{\partial}{\partial t}+\dot{x}\frac{\partial}{\partial x}+\dot{y}\frac{\partial}{\partial y}+\dot{z}\frac{\partial}{\partial z}\right)\varphi_1(\vec{r},t)=\left(Mc^2+eU(\vec{r})\right)\varphi_1(\vec{r},t)$$
$$+c\left(-\mathrm{i}\hbar\frac{\partial}{\partial x}-eA_x(\vec{r},t)\right)\varphi_4(\vec{r},t)-\mathrm{i}c\left(-\mathrm{i}\hbar\frac{\partial}{\partial y}-eA_y(\vec{r},t)\right)\varphi_4(\vec{r},t)+c\left(-\mathrm{i}\hbar\frac{\partial}{\partial z}-eA_z(\vec{r},t)\right)\varphi_3(\vec{r},t) \quad \textbf{(4.88a)}$$

$$-\mathrm{i}\hbar\left(\frac{\partial}{\partial t}+\dot{x}\frac{\partial}{\partial x}+\dot{y}\frac{\partial}{\partial y}+\dot{z}\frac{\partial}{\partial z}\right)\varphi_2(\vec{r},t)=\left(Mc^2+eU(\vec{r})\right)\varphi_2(\vec{r},t)$$
$$+c\left(-\mathrm{i}\hbar\frac{\partial}{\partial x}-eA_x(\vec{r},t)\right)\varphi_3(\vec{r},t)+\mathrm{i}c\left(-\mathrm{i}\hbar\frac{\partial}{\partial y}-eA_y(\vec{r},t)\right)\varphi_3(\vec{r},t)-c\left(-\mathrm{i}\hbar\frac{\partial}{\partial z}-eA_z(\vec{r},t)\right)\varphi_4(\vec{r},t) \quad \textbf{(4.88b)}$$

$$-\mathrm{i}\hbar\left(\frac{\partial}{\partial t}+\dot{x}\frac{\partial}{\partial x}+\dot{y}\frac{\partial}{\partial y}+\dot{z}\frac{\partial}{\partial z}\right)\varphi_3(\vec{r},t)=\left(-Mc^2+eU(\vec{r})\right)\varphi_3(\vec{r},t)$$
$$+c\left(-\mathrm{i}\hbar\frac{\partial}{\partial x}-eA_x(\vec{r},t)\right)\varphi_2(\vec{r},t)-\mathrm{i}c\left(-\mathrm{i}\hbar\frac{\partial}{\partial y}-eA_y(\vec{r},t)\right)\varphi_2(\vec{r},t)+c\left(-\mathrm{i}\hbar\frac{\partial}{\partial z}-eA_z(\vec{r},t)\right)\varphi_1(\vec{r},t) \quad \textbf{(4.88c)}$$

$$-\mathrm{i}\hbar\left(\frac{\partial}{\partial t}+\dot{x}\frac{\partial}{\partial x}+\dot{y}\frac{\partial}{\partial y}+\dot{z}\frac{\partial}{\partial z}\right)\varphi_4(\vec{r},t)=\left(-Mc^2+eU(\vec{r})\right)\varphi_4(\vec{r},t)$$
$$+c\left(-\mathrm{i}\hbar\frac{\partial}{\partial x}-eA_x(\vec{r},t)\right)\varphi_1(\vec{r},t)+\mathrm{i}c\left(-\mathrm{i}\hbar\frac{\partial}{\partial y}-eA_y(\vec{r},t)\right)\varphi_1(\vec{r},t)-c\left(-\mathrm{i}\hbar\frac{\partial}{\partial z}-eA_z(\vec{r},t)\right)\varphi_2(\vec{r},t). \quad \textbf{(4.88d)}$$

To take into account a gravitational field, we generalize the Lagrangian (4.2) for a curved space, with a space-time diagonalization:

$$L\left(x^\alpha,\dot{x}^\alpha\right)=-Mc^2\sqrt{g_{00}+g_{ij}\dot{x}^i\dot{x}^j/c^2}-eU\left(x^i\right)+eA^j\left(x^\mu\right)\dot{x}^j. \quad \textbf{(4.89)}$$

We obtain the momentum

$$P^j=\frac{\partial L}{\partial\dot{x}^j}=\frac{-Mg_{ij}\dot{x}^i}{\sqrt{g_{00}+g_{ik}\dot{x}^i\dot{x}^k/c^2}}+eA^j\left(x^\mu\right), \quad \textbf{(4.90)}$$

and the Hamiltonian

$$H = P^j \dot{x}^j - L\left(x^\alpha, \dot{x}^\alpha\right)$$
$$= \frac{-Mg_{ij}\dot{x}^i\dot{x}^j}{\sqrt{g_{00} + g_{ik}\dot{x}^i\dot{x}^k / c^2}} + \underline{eA^j\left(x^\mu\right)\dot{x}^j} + Mc^2\sqrt{g_{00} + g_{ik}\dot{x}^i\dot{x}^k / c^2} + eU\left(x^i\right) - \underline{eA^j\left(x^\mu\right)\dot{x}^j} \tag{4.91}$$
$$= \frac{Mc^2 g_{00}}{\sqrt{g_{00} + g_{ij}\dot{x}^i\dot{x}^j / c^2}} + eU\left(x^i\right).$$

With the expression

$$\frac{M^2c^2 g_{00}{}^2}{g_{00} + g_{ij}\dot{x}^i\dot{x}^j / c^2} = g_{00}\left(\frac{-M^2 g_{ij}\dot{x}^i\dot{x}^j}{g_{00} + g_{ij}\dot{x}^i\dot{x}^j / c^2} + M^2c^2\right) = g_{00}\left(\frac{M\dot{x}^j\left(P^j - eA^j\right)}{\sqrt{g_{00} + g_{ij}\dot{x}^i\dot{x}^j / c^2}} + M^2c^2\right)$$
$$= g_{00}\left(\delta_i^j \frac{M\dot{x}^i\left(P^j - eA^j\right)}{\sqrt{g_{00} + g_{ij}\dot{x}^i\dot{x}^j / c^2}} + M^2c^2\right) = g_{00}\left(g^{jk} g_{ki} \frac{M\dot{x}^i\left(P^j - eA^j\right)}{\sqrt{g_{00} + g_{ij}\dot{x}^i\dot{x}^j / c^2}} + M^2c^2\right)$$
$$= g_{00}\left[M^2c^2 - g^{jk}\left(P^j - eA^j\right)\left(P^k - eA^k\right)\right],$$

for the Hamiltonian (4.91) we obtain the canonical form

$$H = c\sqrt{g_{00}\left(M^2c^2 - g^{ij}\left[P^i - ecA^i\left(x^\mu\right)\right]\left[P^j - ecA^j\left(x^\mu\right)\right]\right)} + eU\left(x^i\right), \tag{4.92}$$

as a function of the metric elements g_{00} and g^{ij}, which depend on the gravitational potential.

For the interaction with a gravitational wave, we consider the case of a particle with a velocity much smaller than the light velocity. In this case, we can consider the fundamental equation (1.48) of the form

$$\sqrt{g_{00} + g_{ij}\dot{x}^i\dot{x}^j / c^2} \simeq 1. \tag{4.93}$$

as the momentum (4.90) and the Hamiltonian (4.91) take the simpler forms

$$P^j = -Mg_{ij}\dot{x}^i + eA^j\left(x^\mu\right), \tag{4.94}$$

and

$$H = Mc^2 g_{00} + eU\left(x^i\right). \qquad \textbf{(4.95)}$$

By multiplying (4.94) with g^{jk},

$$\begin{aligned} g^{kj}P^j &= -Mg^{jk}g_{ij}\dot{x}^i + eg^{kj}A^j\left(x^\mu\right) = -M\delta_i^k\dot{x}^i + eg^{kj}A^j\left(x^\mu\right) \\ &= -M\dot{x}^k + eg^{kj}A^j\left(x^\mu\right), \end{aligned}$$

we obtain the velocity as a function of the canonical momentum, as a variable of integration in (4.1):

$$\dot{x}^k = \frac{-g^{kj}P^j + eg^{kj}A^j\left(x^\mu\right)}{M}.$$

With these expressions, the Lagrangian (4.89) takes the form

$$\begin{aligned} L\left(x^\alpha, \dot{x}^\alpha\right) &= -Mc^2\sqrt{g_{00} + g_{ij}\dot{x}^i\dot{x}^j / c^2} - eU\left(x^i\right) + eA^j\left(x^\mu\right)\frac{-g^{kj}P^k + eg^{kj}A^k\left(x^\mu\right)}{M} \\ &\simeq -Mc^2\sqrt{g_{00}} - eU\left(x^i\right) - eg^{kj}\frac{P^k}{M}A^j\left(x^\mu\right). \end{aligned} \qquad \textbf{(4.96)}$$

In this expression, we consider a gravitational field with the potential $V_G\left(x^\mu\right)$,

$$g_{00} = 1 + 2V_G\left(x^\mu\right), \qquad \sqrt{g_{00}} \simeq 1 + V_G\left(x^\mu\right) \qquad \textbf{(4.97)}$$

With (4.96) and (4.97), the wave functions (4.1) take the form

$$\psi\left(x^i,t\right)=\frac{1}{\left(2\pi\hbar\right)^{3/2}}\int\varphi\left(P^j,t\right)e^{\frac{\mathrm{i}}{\hbar}\left\{P^j x^j+e\left[U\left(x^i\right)+g^{kj}\frac{P^k}{M}A^j\left(x^\mu\right)+V_G\left(x^\mu\right)\right]t\right\}}e^{\frac{\mathrm{i}}{\hbar}Mc^2 t}\,\mathrm{d}^3P$$

$$\varphi\left(P^j,t\right)=\frac{1}{\left(2\pi\hbar\right)^{3/2}}\int\psi\left(x^i,t\right)e^{-\frac{\mathrm{i}}{\hbar}\left\{P^j x^j+e\left[U\left(x^i\right)+g^{jk}\frac{P^k}{M}A^j\left(x^\mu\right)+V_G\left(x^\mu\right)\right]t\right\}}e^{-\frac{\mathrm{i}}{\hbar}Mc^2 t}\,\mathrm{d}^3x, \quad \textbf{(4.98)}$$

as rapidly varying waves with the frequency proportional to the rest energy, modulated by the electric potential $U\left(x^i\right)$, the radiation vector potential $A^j\left(x^\mu\right)$, and the gravitational potential $V_G\left(x^\mu\right)$. For the general relativistic case, with the Lagrangian (4.89) and the momentum (4.90), the wave functions (4.1) have the explicit form:

$$\psi\left(x^i,t\right)=\frac{1}{\left(2\pi\hbar\right)^{3/2}}\int\varphi\left(P^j,t\right)e^{\frac{\mathrm{i}}{\hbar}\left\{\left(\frac{-Mcg_{ij}\dot{x}^i}{\sqrt{g_{00}+\frac{g_{ik}\dot{x}^i\dot{x}^k}{c^2}}}+eA^j\left(x^\mu\right)\right)x^j+\left[eU\left(x^i\right)-eA^j\left(x^\mu\right)\dot{x}^j+Mc^2\sqrt{g_{00}+\frac{g_{ij}\dot{x}^i\dot{x}^j}{c^2}}\right]t\right\}}\mathrm{d}^3P$$

$$\varphi\left(P^j,t\right)=\frac{1}{\left(2\pi\hbar\right)^{3/2}}\int\psi\left(x^i,t\right)e^{-\frac{\mathrm{i}}{\hbar}\left\{\left(\frac{-Mcg_{ij}\dot{x}^i}{\sqrt{g_{00}+\frac{g_{ik}\dot{x}^i\dot{x}^k}{c^2}}}+eA^j\left(x^\mu\right)\right)x^j+\left[eU\left(x^i\right)-eA^j\left(x^\mu\right)\dot{x}^j+Mc^2\sqrt{g_{00}+\frac{g_{ij}\dot{x}^i\dot{x}^j}{c^2}}\right]t\right\}}\mathrm{d}^3x, \quad \textbf{(4.99)}$$

By a quantum decay $e\Delta U<0$ of this particle in the electric potential $U\left(x^i\right)$, a part of this energy $-e\Delta\tilde{U}>0$ generates a photon with the frequency $\omega=-\frac{e}{\hbar}\Delta\tilde{U}$ and a momentum $eA^i\left(x^\mu\right)$, as the canonical momentum (4.90) is conserved. By the conservation of this momentum,

$$P^j=\frac{-Mg_{ij}\dot{x}^i}{\sqrt{g_{00}+g_{ik}\dot{x}^i\dot{x}^k/c^2}}+eA^j\left(x^\mu\right),$$

from the remaining part of this energy $-e\Delta U-\left(-e\Delta\tilde{U}\right)$ we get a velocity $\dot{x}^i$. In our universe, $r<2m$, this means an increase of the kinetic energy, which with the Schwarzschild metric elements (3.203) is:

$$g_{ij}\dot{x}^i\dot{x}^j=g_{11}\dot{r}^2=\frac{\dot{r}^2}{\frac{2m}{r}-1}. \quad \textbf{(4.100)}$$

This means a red frequency shift $\Delta\omega=\omega-\omega_0=\frac{e}{\hbar}\left(\Delta U-\Delta\tilde{U}\right)<0$, which increases as the distance r approaches the Schwarzschild boundary of our universe, $r\to 2m$.

In the Hamiltonian (4.83) of a free particle,

$$H(\vec{p})=c\sqrt{M^2c^2+\vec{p}^2}, \quad \textbf{(4.101)}$$

we consider the rest momentum Mc as the amplitude of the momentum four-vector p^μ [4],

$$Mc=\sqrt{\left(\frac{H}{c}\right)^2-\vec{p}^2}=\sqrt{p^\mu p_\mu}. \quad \textbf{(4.102)}$$

We consider a four-dimensional square root of this momentum,

$$Mc\hat{\hat{1}}=\sqrt{p^\mu p_\mu}\,\hat{\hat{1}}=\sqrt{g^{\mu\nu}p_\mu p_\nu}\,\hat{\hat{1}}=\gamma^\mu p_\mu, \quad \textbf{(4.103)}$$

depending on the Dirac operators γ^μ. From the square of (4.103), we obtain

$$g^{\mu\nu}p_\mu p_\nu=\gamma^\mu\gamma^\nu p_\mu p_\nu,$$

which, due to the symmetry of the metric tensor, leads to the anticommutation relations:

$$\{\gamma^{\mu},\gamma^{\nu}\}=2g^{\mu\nu}\hat{\hat{1}}. \quad \textbf{(4.104)}$$

For a flat space, with a metric tensor (3.70),

$$g^{00}=1,\ \ g^{11}=-1,\ \ g^{22}=-1,\ \ g^{33}=-1, \quad \textbf{(4.105)}$$

we obtain the squares of these operators

$$\gamma^{0^2}=\hat{\hat{1}},\qquad \gamma^{i^2}=-\hat{\hat{1}}, \quad \textbf{(4.106)}$$

and the anticommutation relations (Clifford algebra):

$$\{\gamma^0,\gamma^i\}=0,\ \ \{\gamma^i,\gamma^j\}_{j\neq i}=0. \quad \textbf{(4.107)}$$

We notice that these anticommutation relations are obtained with the expressions

$$\gamma^0=\alpha_0,\ \ \gamma^i=\alpha_0\alpha_i, \quad \textbf{(4.108)}$$

which, with the first anticommutation relations (4.86), lead to the operator squares (4.106),

$$\gamma^{0^2}=\alpha_0{}^2=\hat{\hat{1}},\qquad \gamma^{i^2}=\alpha_0\alpha_i\alpha_0\alpha_i=-\alpha_0\alpha_0\alpha_i\alpha_i=-\hat{\hat{1}}$$

and the anticommutation relations (4.107):

$$\{\gamma^0,\gamma^i\}=\{\alpha_0,\alpha_0\alpha_i\}=\alpha_0\alpha_0\alpha_i+\alpha_0\alpha_i\alpha_0=\alpha_0\alpha_0\alpha-\alpha_0\alpha_0\alpha=0,$$
$$\{\gamma^i,\gamma^j\}_{j\neq i}=\{\alpha_0\alpha_i,\alpha_0\alpha_j\}=\alpha_0\alpha_i\alpha_0\alpha_j+\alpha_0\alpha_j\alpha_0\alpha_i=-\alpha_0{}^2\{\alpha_i,\alpha_j\}=0.$$

Thus, from (4.108) with (4.84), we obtain the Dirac operators

$$\gamma^0=\alpha_0=\begin{pmatrix}\hat{1} & \hat{0}\\ \hat{0} & -\hat{1}\end{pmatrix},\quad \gamma^i=\alpha_0\alpha_i=\begin{pmatrix}\hat{1} & \hat{0}\\ \hat{0} & -\hat{1}\end{pmatrix}\begin{pmatrix}\hat{0} & \sigma_i\\ \sigma_i & \hat{0}\end{pmatrix}=\begin{pmatrix}0 & \sigma_i\\ -\sigma_i & 0\end{pmatrix}. \quad \textbf{(4.109)}$$

In the two dynamic equations (4.82) for the time-dependent wave four-vectors, we consider Dirac's Hamiltonian of the forms

$$H\hat{\hat{1}} = c\sqrt{M^2c^2 + \vec{p}^2}\,\hat{\hat{1}} = c\left(\alpha_0 Mc + \alpha_1 p_1 + \alpha_2 p_2 + \alpha_3 p_3\right),$$

and respectively,

$$H\hat{\hat{1}} = c\sqrt{M^2c^2 + \vec{p}^2}\,\hat{\hat{1}} = c\left(\alpha_0 Mc + \alpha_1 p^1 + \alpha_2 p^2 + \alpha_3 p^3\right).$$

We obtain

$$\begin{aligned} i\hbar\frac{\partial}{\partial t}\psi_t(\vec{r},t)\hat{\hat{1}} &= \left[\vec{p}\dot{\vec{r}}\hat{\hat{1}} - c\left(\alpha_0 Mc + \alpha_1 p_1 + \alpha_2 p_2 + \alpha_3 p_3\right)\right]\psi_t(\vec{r},t) \\ i\hbar\frac{\partial}{\partial t}\varphi_t(\vec{p},t)\hat{\hat{1}} &= -\left[\vec{p}\dot{\vec{r}}\hat{\hat{1}} - c\left(\alpha_0 Mc + \alpha_1 p^1 + \alpha_2 p^2 + \alpha_3 p^3\right)\right]\varphi_t(\vec{p},t). \end{aligned} \quad \textbf{(4.110)}$$

With the momentum operators,

$$p_0 = p^0 = i\hbar\frac{\partial}{\partial x^0} = i\hbar\frac{\partial}{c\partial t}, \qquad p_i = -p^i = i\hbar\frac{\partial}{\partial x^i}, \quad \textbf{(4.111)}$$

equations (4.110) become:

$$\begin{aligned} &\left[i\hbar\left(\hat{\hat{1}}\frac{\partial}{\partial x^0} + \alpha_1\frac{\partial}{\partial x^1} + \alpha_2\frac{\partial}{\partial x^2} + \alpha_3\frac{\partial}{\partial x^3}\right) + \alpha_0 Mc - \vec{p}\frac{\dot{\vec{r}}}{c}\hat{\hat{1}}\right]\psi_t(\vec{r},t) = 0 \\ &\left(p^0 - \alpha_1 p^1 - \alpha_2 p^2 - \alpha_3 p^3 - \alpha_0 Mc + \vec{p}\frac{\dot{\vec{r}}}{c}\hat{\hat{1}}\right)\varphi_t(\vec{p},t) = 0. \end{aligned}$$

By multiplying with $\alpha_0 = \gamma^0$, in these equations, we obtain the Dirac matrices (4.108) as coefficients. With the momentum operators (4.111), these equations take the form

$$\left[i\hbar\left(\gamma^0\frac{\partial}{\partial x^0}+\gamma^1\frac{\partial}{\partial x^1}+\gamma^2\frac{\partial}{\partial x^2}+\gamma^3\frac{\partial}{\partial x^3}\right)+\hat{\hat{1}}Mc-\gamma^0\vec{p}\frac{\dot{\vec{r}}}{c}\right]\psi_t\left(\vec{r},t\right)=0$$
$$\left(\gamma^0 p_0+\gamma^1 p_1+\gamma^2 p_2+\gamma^3 p_3-\hat{\hat{1}}Mc+\gamma^0\vec{p}\frac{\dot{\vec{r}}}{c}\right)\varphi_t\left(\vec{p},t\right)=0. \tag{4.112}$$

From the relativistic mechanical momentum

$$\vec{p}=\frac{M\dot{\vec{r}}}{\sqrt{1-\frac{\dot{\vec{r}}^2}{c^2}}},$$

we obtain the normalized velocity

$$\frac{\dot{\vec{r}}}{c}=\frac{\vec{p}}{\sqrt{M^2c^2+\vec{p}^2}}=\frac{\vec{p}}{E}, \tag{4.113}$$

as a function of this momentum and the normalized energy

$$E=\sqrt{M^2c^2+\vec{p}^2}. \tag{4.114}$$

With this expression, we obtain the term

$$\vec{p}\frac{\dot{\vec{r}}}{c}=\frac{\vec{p}^2}{E}.$$

With the rest momentum

$$\mu=Mc, \tag{4.115}$$

and the normalized momentum

$$\eta=\frac{\vec{p}^2}{\mu E}=\frac{\vec{p}^2}{\mu\sqrt{\mu^2+\vec{p}^2}}\in\left(0,\infty\right), \tag{4.116}$$

equations (4.112) take the form

$$\begin{aligned}&\left[\mathrm{i}\hbar\left(\gamma^0\frac{\partial}{\partial x^0}+\gamma^1\frac{\partial}{\partial x^1}+\gamma^2\frac{\partial}{\partial x^2}+\gamma^3\frac{\partial}{\partial x^3}\right)+\mu\left(\hat{\hat{1}}-\gamma^0\eta\right)\right]\psi_t(\vec{r},t)=0\\&\left[\gamma^0 p_0+\gamma^1 p_1+\gamma^2 p_2+\gamma^3 p_3-\mu\left(\hat{\hat{1}}-\gamma^0\eta\right)\right]\varphi_t(\vec{p},t)=0.\end{aligned} \quad \textbf{(4.117)}$$

With the notations

$$\begin{aligned}&\not{\partial}=\gamma^0\frac{\partial}{\partial x^0}+\gamma^1\frac{\partial}{\partial x^1}+\gamma^2\frac{\partial}{\partial x^2}+\gamma^3\frac{\partial}{\partial x^3}=\gamma^\mu\partial_\mu\\&\not{p}=\gamma^0 p_0+\gamma^1 p_1+\gamma^2 p_2+\gamma^3 p_3=\gamma^\mu p_\mu,\end{aligned} \quad \textbf{(4.118)}$$

these equations take the shorter form

$$\begin{aligned}&\left[-\mathrm{i}\hbar\not{\partial}-\mu\left(\hat{\hat{1}}-\gamma^0\eta\right)\right]\psi_t(x)=0\\&\left[\not{p}-\mu\left(\hat{\hat{1}}-\gamma^0\eta\right)\right]\varphi_t(p)=0,\end{aligned} \quad \textbf{(4.119)}$$

which can be compared to the similar equations in [4]. With a solution of the form

$$\varphi_t(p)\sim u(p)=\begin{pmatrix}\tilde{u}(p)\\ \tilde{v}(p)\end{pmatrix}=\begin{pmatrix}u_1(p)\\ u_2(p)\\ u_3(p)\\ u_4(p)\end{pmatrix}, \quad \textbf{(4.120)}$$

the second equation (4.119) is

$$\left[\gamma^0 p_0 + \gamma^1 p_1 + \gamma^2 p_2 + \gamma^3 p_3 - \mu\left(\hat{\hat{1}} - \gamma^0 \eta\right)\right]\begin{pmatrix} u_1(p) \\ u_2(p) \\ u_3(p) \\ u_4(p) \end{pmatrix} = 0. \tag{4.121}$$

For this equation with the matrix elements (4.109),

$$\gamma^0 = \begin{pmatrix} 1 & 0 & 0 & 0 \\ 0 & 1 & 0 & 0 \\ 0 & 0 & -1 & 0 \\ 0 & 0 & 0 & -1 \end{pmatrix} \quad \gamma^1 = \begin{pmatrix} 0 & 0 & 0 & 1 \\ 0 & 0 & 1 & 0 \\ 0 & -1 & 0 & 0 \\ -1 & 0 & 0 & 0 \end{pmatrix}$$
$$\gamma^2 = \begin{pmatrix} 0 & 0 & 0 & -\mathrm{i} \\ 0 & 0 & \mathrm{i} & 0 \\ 0 & \mathrm{i} & 0 & 0 \\ -\mathrm{i} & 0 & 0 & 0 \end{pmatrix} \quad \gamma^3 = \begin{pmatrix} 0 & 0 & 1 & 0 \\ 0 & 0 & 0 & -1 \\ -1 & 0 & 0 & 0 \\ 0 & 1 & 0 & 0 \end{pmatrix}, \tag{4.122}$$

we consider a reference system with the z -axis in the direction of the momentum component $p_3 = |\vec{p}|$, $p_1 = p_2 = 0$. In this case, equation (4.121) takes the explicit form

$$\begin{pmatrix} p_0 + \mu\eta - \mu & 0 & p_3 & 0 \\ 0 & p_0 + \mu\eta - \mu & 0 & -p_3 \\ -p_3 & 0 & -(p_0 + \mu\eta + \mu) & 0 \\ 0 & p_3 & 0 & -(p_0 + \mu\eta + \mu) \end{pmatrix}\begin{pmatrix} u_1(p) \\ u_2(p) \\ u_3(p) \\ u_4(p) \end{pmatrix} = 0 \tag{4.123}$$

For a nontrivial solution of this homogeneous system of equations, its determinant must be null:

$$
\begin{vmatrix}
p_0-\mu(1-\eta) & 0 & p_3 & 0 \\
0 & p_0-\mu(1-\eta) & 0 & -p_3 \\
-p_3 & 0 & -[p_0+\mu(1+\eta)] & 0 \\
0 & p_3 & 0 & -[p_0+\mu(1+\eta)]
\end{vmatrix}
$$

$$
=[p_0-\mu(1-\eta)]\{[p_0-\mu(1-\eta)][p_0+\mu(1+\eta)]^2-p_3^2[p_0+\mu(1+\eta)]\}
$$

$$
+p_3[p_3^3-[p_0+\mu(1+\eta)][p_0-\mu(1-\eta)]p_3]
$$

$$
=[(p_0+\mu\eta)^2-p_3^2-\mu^2]^2=0.
$$

From this equation, we obtain the expression

$$
p_0+\mu\eta=p_0+\frac{\vec{p}\dot{\vec{r}}}{c}=\pm E\,, \tag{4.124}
$$

of the normalized energy as a function of the momentum,

$$
E=\sqrt{p_3^2+\mu^2}=\sqrt{\vec{p}^2+\mu^2}\,. \tag{4.125}
$$

From (123) for a positive normalized energy $+E$, for the two two-dimensional wave functions (4.120), $\tilde{u}_+(p)$ and $\tilde{v}_+(p)$, we obtain the dynamic equations

$$
\begin{aligned}
&(E-\mu)\tilde{u}_+(p)+\sigma_3 p_3\tilde{v}_+(p)=0 \\
&\sigma_3 p_3\tilde{u}_+(p)+(E+\mu)\tilde{v}_+(p)=0.
\end{aligned} \tag{4.126}
$$

In these equations, we consider the two-dimensional wave function $\tilde{u}_+(p)$ as the spin wave function, with the spin eigenfunctions

$$
\tilde{u}_+(p)=\begin{pmatrix}1\\0\end{pmatrix},\ \begin{pmatrix}0\\1\end{pmatrix}, \tag{4.127}
$$

and the normalization condition

$$\tilde{u}_+\dagger(p)\tilde{u}_+(p)=1. \tag{4.128}$$

From the second equation (4.126), we obtain

$$\begin{aligned}\tilde{v}_+(p)&=-\frac{\sigma_3 p_3}{E+\mu}\tilde{u}_+(p)=\frac{\sigma_3 p^3}{E+\mu}\tilde{u}_+(p)\\&=\frac{\vec{\sigma}\vec{p}}{E+\mu}\tilde{u}_+(p).\end{aligned} \tag{4.129}$$

This solution is obtained also from the first equation (4.126), by multiplying with the Pauli matrix σ_3 with the normalization relation $\sigma_3^{\ 2}=\hat{1}$, and by using the relation (4.125):

$$\begin{aligned}\tilde{v}_+(p)&=-\frac{E-\mu}{p_3}\sigma_3\tilde{u}_+(p)=-\frac{(E-\mu)(E+\mu)}{p_3(E+\mu)}\sigma_3\tilde{u}_+(p)=-\frac{\sigma_3 p_3}{E+\mu}\tilde{u}_+(p)\\&=\frac{\vec{\sigma}\vec{p}}{E+\mu}\tilde{u}_+(p)\end{aligned}.$$

With these expressions for a positive normalized energy E, the wave function (4.120) of a particle is

$$u_+(p)=\begin{pmatrix}\tilde{u}_+(p)\\ \dfrac{\vec{\sigma}\vec{p}}{E+\mu}\tilde{u}_+(p)\end{pmatrix}. \tag{4.130}$$

For a negative normalized energy $-E$, for the two-dimensional wave functions (4.120) with the comonents $\tilde{u}_-(p)$ and $\tilde{v}_-(p)$, from (4.123) we obtain the dynamic equations

$$\begin{aligned}&-(E+\mu)\tilde{u}_-(p)+\sigma_3 p_3\tilde{v}_-(p)=0\\&\sigma_3 p_3\tilde{u}_-(p)-(E-\mu)\tilde{v}_-(p)=0,\end{aligned} \tag{4.131}$$

where we consider the normalized spin wave function $\tilde{v}_-(p)$, with the spin eigenfunctions

$$\tilde{v}_-(p) = \begin{pmatrix} 1 \\ 0 \end{pmatrix}, \begin{pmatrix} 0 \\ 1 \end{pmatrix}, \qquad \textbf{(4.132)}$$

and the normalization relation

$$\tilde{v}_-^\dagger(p)\tilde{v}_-(p) = 1. \qquad \textbf{(4.133)}$$

From the first equation (4.131), we obtain

$$\begin{aligned} \tilde{u}_-(p) &= \frac{\sigma_3 p_3}{E+\mu}\tilde{v}_-(p) = -\frac{\sigma_3 p^3}{E+\mu}\tilde{v}_-(p) \\ &= -\frac{\vec{\sigma}\vec{p}}{E+\mu}\tilde{v}_-(p). \end{aligned} \qquad \textbf{(4.134)}$$

The same solution is also obtained from the second equation (4.131):

$$\begin{aligned} \tilde{u}_-(p) &= \frac{E-\mu}{p_3}\sigma_3\tilde{v}_-(p) = \frac{(E-\mu)(E+\mu)}{p_3(E+\mu)} = \frac{\sigma_3 p_3}{E+\mu}\tilde{v}_-(p) \\ &= -\frac{\vec{\sigma}\vec{p}}{E+\mu}\tilde{v}_-(p) \end{aligned}$$

With these expressions, for negative normalized energy $-E$ as a characteristic of an antiparticle, the wave function (4.120) is

$$u_-(p) = \begin{pmatrix} -\dfrac{\vec{\sigma}\vec{p}}{E+\mu}\tilde{v}_-(p) \\ \tilde{v}_-(p) \end{pmatrix}. \qquad \textbf{(4.135)}$$

With the wave functions (4.130) and (4.135), we obtain the wave function of a particle-antiparticle system with the particle and antiparticle amplitudes, α and respectively β:

$$u(p)=\alpha u_+(p)+\beta u_-(p)=\alpha\begin{pmatrix}\tilde{u}_+(p)\\ \dfrac{\vec{\sigma}\vec{p}}{E+\mu}\tilde{u}_+(p)\end{pmatrix}+\beta\begin{pmatrix}-\dfrac{\vec{\sigma}\vec{p}}{E+\mu}\tilde{v}_-(p)\\ \tilde{v}_-(p)\end{pmatrix}. \quad \textbf{(4.136)}$$

To obtain the normalization condition for these amplitudes, we calculate the lengths of the four-vectors (4.130) and (4.135). For a particle four-vector, we obtain the length

$$\begin{aligned}u_+^\dagger(p)u_+(p)&=\begin{pmatrix}\tilde{u}_+^\dagger(p) & \tilde{u}_+^\dagger(p)\dfrac{\vec{\sigma}\vec{p}}{E+\mu}\end{pmatrix}\begin{pmatrix}\tilde{u}_+(p)\\ \dfrac{\vec{\sigma}\vec{p}}{E+\mu}\tilde{u}_+(p)\end{pmatrix}\\ &=\tilde{u}_+^\dagger(p)\tilde{u}_+(p)+\tilde{u}_+^\dagger(p)\frac{\vec{\sigma}\vec{p}}{E+\mu}\frac{\vec{\sigma}\vec{p}}{E+\mu}\tilde{u}_+(p)\\ &=\left(1+\frac{E^2-\mu^2}{(E+\mu)^2}\right)\tilde{u}_+^\dagger(p)\tilde{u}_+(p)\\ &=\left(1+\frac{E-\mu}{E+\mu}\right)\tilde{u}_+^\dagger(p)\tilde{u}_+(p)=\frac{2E}{E+\mu},\end{aligned} \quad \textbf{(4.137)}$$

and the same length of an antiparticle four-vector,

$$\begin{aligned}u_-^\dagger(p)u_-(p)&=\begin{pmatrix}-\tilde{v}_-^\dagger(p)\dfrac{\vec{\sigma}\vec{p}}{E+\mu} & \tilde{v}_-^\dagger(p)\end{pmatrix}\begin{pmatrix}-\dfrac{\vec{\sigma}\vec{p}}{E+\mu}\tilde{v}_-(p)\\ \tilde{v}_-(p)\end{pmatrix}\\ &=\tilde{v}_-^\dagger(p)\frac{\vec{\sigma}\vec{p}}{E+\mu}\frac{\vec{\sigma}\vec{p}}{E+\mu}\tilde{v}_-(p)+\tilde{v}_-^\dagger(p)\tilde{v}_-(p)\\ &=\left(\frac{E^2-\mu^2}{(E+\mu)^2}+1\right)\tilde{v}_-^\dagger(p)\tilde{v}_-(p)\\ &=\left(\frac{E-\mu}{E+\mu}+1\right)\tilde{v}_-^\dagger(p)\tilde{v}_-(p)=\frac{2E}{E+\mu}.\end{aligned} \quad \textbf{(4.138)}$$

With these expressions, from the condition of a covariant normalization of the wave function (4.136) [4], we obtain the relation

$$u^{\dagger}(p)u(p)=|\alpha|^{2}u_{+}^{\dagger}(p)u_{+}(p)+|\beta|^{2}u_{-}^{\dagger}(p)u_{-}(p)$$
$$=\left(|\alpha|^{2}+|\beta|^{2}\right)\frac{2E}{E+\mu}=\frac{E}{\mu},$$

which leads to the normalization condition

$$|\alpha|^{2}+|\beta|^{2}=\frac{E+\mu}{2\mu}. \qquad \textbf{(4.139)}$$

For the solution (4.120) of the second equation (4.119), we consider an expression of the form

$$\varphi_{t}(p)=\frac{1}{L_{p}^{3/2}}u(p), \qquad \textbf{(4.140)}$$

with the normalization length L_{p}, which will be determined from the normalization condition of the propagation function (4.81). With (4.140), the second wave function (4.81), of the particle in the momentum space becomes

$$\varphi(\vec{p},t)=e^{-\frac{i}{\hbar}\vec{p}\hat{\vec{r}}}\varphi_{t}(p)=\frac{1}{L_{p}^{3/2}}e^{-\frac{i}{\hbar}\vec{p}\vec{r}}u(p)=\varphi(\vec{p}). \qquad \textbf{(4.141)}$$

With this wave function in the momentum space, we obtain the first wave equation (4.81) of the particle in the coordinate space, with dimensions determined by the domain $\Delta^{3}\vec{p}$ of the momentum, $\vec{p}\in\Delta^{3}\vec{p}$,

$$\psi(\vec{r},t)=\frac{1}{\left(2\pi\hbar L_p\right)^{3/2}}\int_{\Delta^3\vec{p}}u(p)e^{-\frac{i}{\hbar}\left[\vec{p}\dot{\vec{r}}-H(\vec{p},\vec{r})\right]t}\mathrm{d}^3\vec{p}. \quad \textbf{(4.142)}$$

With the wave function (4.136) of a free particle-antiparticle system, $\vec{r}=\dot{\vec{r}}t$, this wave function is of the form

$$\psi(\vec{r},t)=\frac{1}{\left(2\pi\hbar L_p\right)^{3/2}}\int_{\Delta^3\vec{p}}e^{-\frac{i}{\hbar}\vec{p}\vec{r}}\left[\alpha u_+(p)e^{\frac{i}{\hbar}cEt}+\beta u_-(p)e^{-\frac{i}{\hbar}cEt}\right]\mathrm{d}^3\vec{p}, \quad \textbf{(4.143)}$$

with the particle and antiparticle amplitudes α and β which satisfy the normalization condition (4.139). With this normalization condition, and the normalization conditions (4.137) and (4.138), we find the normalization condition of this wave function:

$$\begin{aligned} L_p^{\,3} &= \int_{\Delta^3\vec{p}}\frac{2E}{E+\mu}\left(|\alpha|^2+|\beta|^2\right)\mathrm{d}^3\vec{p}=\int_{\Delta^3\vec{p}}\frac{2E}{E+\mu}\frac{E+\mu}{2\mu}\mathrm{d}^3\vec{p}=\int_{\Delta^3\vec{p}}\frac{E}{\mu}\mathrm{d}^3\vec{p} \\ &= \int_{\Delta^3\vec{p}}\sqrt{1+\frac{\vec{p}^{\,2}}{\mu^2}}\mathrm{d}^3\vec{p}, \end{aligned} \quad \textbf{(4.144)}$$

in agreement with the nonrelativistic case $L_p^{\,3}=\Delta^3\vec{p}$. The wave function (4.143) has been obtained as a solution of the second equation (4.119). This equation can be obtained also as a solution of the first equation (4.119), which is the inverse transform of the second equation. For this, we consider the relation

$$\psi_t(x)=e^{\frac{i}{\hbar}px}\varphi_t(p)=\frac{1}{L_p^{\,3/2}}e^{\frac{i}{\hbar}px}u(p), \quad \textbf{(4.145)}$$

as a function of the momentum-coordinate scalar product

$$\begin{aligned} px &= p_0 x^0 + p_1 x^1 + p_2 x^2 + p_3 x^3 \\ &= p_0 x^0 - \vec{p}\vec{r}\,. \end{aligned} \tag{4.146}$$

With this relation, the first term of the first equation (4.119) takes the form of the first term of the second equation (4.119)

$$\begin{aligned} -\mathrm{i}\hbar\not{\partial} e^{\frac{\mathrm{i}}{\hbar}px} &= \left(\gamma^0 p_0 + \gamma^1 p_1 + \gamma^2 p_2 + \gamma^3 p_3\right) e^{\frac{\mathrm{i}}{\hbar}px} \\ &= \not{p} e^{\frac{\mathrm{i}}{\hbar}px}\,, \end{aligned}$$

which means that the two equations are equivalent. From the first wave packet (4.81), of momentum eigenfunctions serving (4.140) as solutions of the first equation (4.119), we obtain the wave function

$$\begin{aligned} \psi(\vec{r},t) = e^{\frac{\mathrm{i}}{\hbar}\hat{\vec{p}}\vec{r}} \frac{1}{\left(2\pi\hbar L_p\right)^{3/2}} \int e^{\frac{\mathrm{i}}{\hbar}px} u(p)\,\mathrm{d}^3\vec{p} &= \frac{1}{\left(2\pi\hbar L_p\right)^{3/2}} \int e^{\frac{\mathrm{i}}{\hbar}p_0 x^0} u(p)\,\mathrm{d}^3\vec{p} \\ &= \frac{1}{\left(2\pi\hbar L_p\right)^{3/2}} \int e^{-\frac{\mathrm{i}}{\hbar}\vec{p}\vec{r}} e^{\frac{\mathrm{i}}{\hbar}\left(p_0 x^0 + \frac{\vec{p}\dot{\vec{r}}}{c} ct\right)} u(p)\,\mathrm{d}^3\vec{p} \\ &= \frac{1}{\left(2\pi\hbar L_p\right)^{3/2}} \int_{\Delta^3\vec{p}} e^{-\frac{\mathrm{i}}{\hbar}\vec{p}\vec{r}} \left[\alpha u_+(p) e^{\frac{\mathrm{i}}{\hbar}cEt} + \beta u_-(p) e^{-\frac{\mathrm{i}}{\hbar}cEt}\right] \mathrm{d}^3\vec{p}\,, \end{aligned}$$

of the form (143) obtained from the second equation (4.119). From these wave packets with (4.113), we obtain the group velocity,

$$\begin{aligned} \frac{\partial}{\partial\vec{p}} cE &= \frac{\partial}{\partial\vec{p}} c\sqrt{M^2c^2 + \vec{p}^2} \\ &= \frac{c\vec{p}}{\sqrt{M^2c^2 + \vec{p}^2}} = \dot{\vec{r}}\,, \end{aligned} \tag{4.147}$$

equal to the velocity of the matter described by this wave packet. In the expression (4.130) we distinguish two two-dimensional components: the spin component $\tilde{u}_+(p)$, with an amplitude $\tilde{u}_+ \dagger(p)\tilde{u}_+(p)=1$, and the momentum-spin component $\frac{\vec{\sigma}\vec{p}}{E+\mu}\tilde{u}_+(p)$ with the amplitude

$$\tilde{u}_+^\dagger(p)\frac{\vec{\sigma}\vec{p}}{E+\mu}\frac{\vec{\sigma}\vec{p}}{E+\mu}\tilde{u}_+(p)=\frac{E-\mu}{E+\mu}\tilde{u}_+^\dagger(p)\tilde{u}_+(p)=\frac{E-\mu}{E+\mu}<1, \qquad \textbf{(4.148)}$$

tending to 1 in the ultra-relativistic case.

CONCLUSION

We described the dynamics of a quantum particle in an electromagnetic field, by additional Lagrangian terms in the time-dependent phases of the two conjugate wave functions: a term conjugated to time with electric potential $U(\vec{r})$, and a term conjugated to the coordinates, with a vector potential $\vec{A}(\vec{r},t)$. From the group velocity in the momentum space, we obtained the Lorentz force, as a function of the electric and magnetic fields satisfying three of the four Maxwell equations. From the physical condition of a field propagation velocity equal to the maximum relativistic velocity c, we obtained the fourth Maxwell equation and the conservation condition of the electric charge. For a quantum particle in a gravitational field, we considered a metric tensor proportional to an amplitude tensor, and a polarization vector in the first-order approximation, or a polarization two-dimensional tensor in the second-order approximation. For the amplitude tensor, we found a rotation in a plane perpendicular to the propagation direction of the gravitational wave with a spin 2, which we call the graviton spin, and a rotation of the polarization vector in this plane, which we call the particle spin, with a half-integer value for Fermions and an integer value for Bosons. We considered the propagation wave functions as products of propagation factors with the time-dependent wave-functions, which satisfy time-dependent equations similar to the Schrödinger equation. For a particle state four-vector, we found Dirac-like equations, with additional terms depending on velocity. For a particle decay in electric potential, a redshift of the generated electromagnetic field has been obtained due to the gravitational field. From the time-dependent dynamic equations for a free particle, we obtained equations depending on the Clifford elements, which compared to the similar equations of the quantum field theory, include additional, momentum dependent terms for their rest mass dependent eigenvalues. The solution of these equations for particles and antiparticles, is a wave packet over the momentum spectrum, as a finite invariant distribution of matter propagating in space.

Open Quantum Physics, 2022, *Vol. 3*, 130-153

CHAPTER 5

The Least Action and Matter-Field Dynamics in Gravitational Field

Abstract: We define the gravitation action as an integral over the four-dimensional space of the total curvature density. Integrating by parts, we obtain the Lagrangian density as a function of the Christoffel symbols. From a variation of this Lagrangian density with the metric elements, we obtain the Einstein law of gravitation for vacuum. We define the matter action as an integral over the four-dimensional space of the mass scalar density. With the momentum four-vector, we obtain a Lagrangian density variation with the metric elements and the time-space coordinates. From the total variation of the matter action in a gravitational field, we obtain the Einstein law of gravitation in matter, and the geodesic equations. According to the Maxwell equations, we obtain the electric and magnetic fields as an electromagnetic tensor. We define the electromagnetic action as an integral over the four-dimensional space of the amplitude scalar of this tensor, and obtain a variation of this action with metric elements and the electromagnetic potentials. At the same time, we consider the electric charge action as the scalar product of the charge flux four-vector with the electromagnetic potential four-vector, and obtain its variation with electromagnetic potentials and the time-space coordinates. We obtain a matter-field action variation with three terms: for the variations of the metric elements, of the time-space coordinates, and of the electromagnetic potentials. From the first term, we obtain the matter dynamics in a gravitational-electromagnetic field; from the second term, we obtain the Lorentz force in a gravitational field, and from the third term, the Maxwell equations in the gravitational field.

Keywords: Action, Curvature tensor, Ricci tensor, Christoffel symbol, Metric tensor, Lagrangian density, Variation, Fundamental equation, Covariant acceleration, Geodesic equation, Maxwell equations, Momentum four-vector, Electric field, Magnetic field, Electromagnetic potential, Electric potential, Pseudo-energy tensor, Vector potential, Electromagnetic field tensor, Charge density, Current density, Scalar density, Electromagnetic energy density, Poynting vector, Strength energy tensor.

5.1. GRAVITATION ACTION

We consider the action of the gravitational field as the integral of the scalar curvature over the four-dimensional physical space,

Eliade Stefanescu

$$I = \int R\sqrt{-g}\mathrm{d}^4 x, \tag{5.1}$$

obtained by a double contraction of the four-dimensional curvature tensor (3.100),

$$R^{\alpha}_{\mu\nu\sigma} = -\Gamma^{\alpha}_{\mu\nu,\sigma} - \Gamma^{\beta}_{\mu\nu}\Gamma^{\alpha}_{\beta\sigma} + \Gamma^{\alpha}_{\mu\sigma,\nu} + \Gamma^{\beta}_{\mu\sigma}\Gamma^{\alpha}_{\beta\nu}. \tag{5.2}$$

By the first contraction, $\alpha = \sigma$, we obtain the Ricci symmetric tensor (3.124),

$$R_{\mu\nu} = R_{\nu\mu} = R^{\alpha}_{\mu\nu\alpha} = -\Gamma^{\alpha}_{\mu\nu,\alpha} - \Gamma^{\beta}_{\mu\nu}\Gamma^{\alpha}_{\beta\alpha} + \Gamma^{\alpha}_{\mu\alpha,\nu} + \Gamma^{\beta}_{\mu\alpha}\Gamma^{\alpha}_{\beta\nu}. \tag{5.3}$$

The symmetry of the Ricci tensor is obtained from the symmetry of the Christoffel symbol $\Gamma^{\alpha}_{\nu\rho} = \Gamma^{\alpha}_{\rho\nu}$, and the expression (3.66). With the symmetric tensors (5.3) and $g^{\mu\nu}$, we perform the second contraction, which leads to the scalar curvature (3.121):

$$R = g^{\mu\nu} R_{\mu\nu}. \tag{5.4}$$

We consider (5.1) with (5.4)-(5.3) as an integral of the difference

$$R = R^* - L, \tag{5.5}$$

between the first-order term

$$R^* = g^{\mu\nu}\left(\Gamma^{\alpha}_{\mu\alpha,\nu} - \Gamma^{\alpha}_{\mu\nu,\alpha}\right), \tag{5.6}$$

and the second-order term

$$L = g^{\mu\nu}\left(\Gamma^{\beta}_{\mu\nu}\Gamma^{\alpha}_{\beta\alpha} - \Gamma^{\beta}_{\mu\alpha}\Gamma^{\alpha}_{\beta\nu}\right). \tag{5.7}$$

For the term (5.6), we consider a partial integration,

$$
\begin{aligned}
R^{*}\sqrt{-g} &= g^{\mu\nu}\Gamma^{\alpha}_{\mu\alpha,\nu}\sqrt{-g} - g^{\mu\nu}\Gamma^{\alpha}_{\mu\nu,\alpha}\sqrt{-g} \\
&= \left(g^{\mu\nu}\Gamma^{\alpha}_{\mu\alpha}\sqrt{-g}\right)_{,\nu} - \Gamma^{\alpha}_{\mu\alpha}\left(g^{\mu\nu}\sqrt{-g}\right)_{,\nu} - \left(g^{\mu\nu}\Gamma^{\alpha}_{\mu\nu}\sqrt{-g}\right)_{,\alpha} + \Gamma^{\alpha}_{\mu\nu}\left(g^{\mu\nu}\sqrt{-g}\right)_{,\alpha}
\end{aligned}
$$

with the perfect differentials having no contribution in the action integral (5.1), so that we are left with

$$
R^{*}\sqrt{-g} = \Gamma^{\alpha}_{\mu\nu}\left(g^{\mu\nu}\sqrt{-g}\right)_{,\alpha} - \Gamma^{\alpha}_{\mu\alpha}\left(g^{\mu\nu}\sqrt{-g}\right)_{,\nu} . \qquad \textbf{(5.8)}
$$

With (3.66),

$$
\sqrt{-g}_{,\nu} = \Gamma^{\mu}_{\nu\mu}\sqrt{-g} , \qquad \textbf{(5.9)}
$$

(3.46), and (3.41), we obtain the coordinate derivative in the first term of (5.8),

$$
\begin{aligned}
\left(g^{\mu\nu}\sqrt{-g}\right)_{,\alpha} &= g^{\mu\nu}{}_{,\alpha}\sqrt{-g} + g^{\mu\nu}\sqrt{-g}_{,\alpha} = -g^{\mu\rho}g^{\nu\sigma}g_{\rho\sigma,\alpha}\sqrt{-g} + g^{\mu\nu}\Gamma^{\sigma}_{\alpha\sigma}\sqrt{-g} \\
&= -g^{\mu\rho}g^{\nu\sigma}\left(\Gamma_{\rho\sigma\alpha} + \Gamma_{\sigma\rho\alpha}\right)\sqrt{-g} + g^{\mu\nu}\Gamma^{\sigma}_{\alpha\sigma}\sqrt{-g} \\
&= \left(-g^{\nu\sigma}\Gamma^{\mu}_{\sigma\alpha} - g^{\mu\rho}\Gamma^{\nu}_{\rho\alpha} + g^{\mu\nu}\Gamma^{\sigma}_{\alpha\sigma}\right)\sqrt{-g}.
\end{aligned} \qquad \textbf{(5.10)}
$$

By the contraction $\alpha = \nu$, this coordinate derivative takes the form of the coordinate derivative in the second term,

$$
\left(g^{\mu\nu}\sqrt{-g}\right)_{,\nu} = \left(-g^{\nu\sigma}\Gamma^{\mu}_{\sigma\nu} \underline{- g^{\mu\rho}\Gamma^{\nu}_{\rho\nu}} + \underline{g^{\mu\nu}\Gamma^{\sigma}_{\nu\sigma}}\right)\sqrt{-g} = -g^{\nu\sigma}\Gamma^{\mu}_{\sigma\nu}\sqrt{-g} . \qquad \textbf{(5.11)}
$$

Due to the symmetry of the Christoffel symbol, $\Gamma^{\alpha}_{\mu\nu} = \Gamma^{\alpha}_{\nu\mu}$, with (5.10) and (5.11), we find that the term R^{*} is double of the term (5.7):

$$
\begin{aligned}
R^*\sqrt{-g} &= \left(\underline{-g^{\nu\sigma}\Gamma^{\mu}_{\sigma\alpha} - g^{\mu\rho}\Gamma^{\nu}_{\rho\alpha}} + g^{\mu\nu}\Gamma^{\sigma}_{\alpha\sigma}\right)\Gamma^{\alpha}_{\mu\nu}\sqrt{-g} + \underline{\underline{g^{\nu\sigma}\Gamma^{\mu}_{\sigma\nu}\Gamma^{\alpha}_{\mu\alpha}}}\sqrt{-g} \\
&= 2\left(g^{\mu\nu}\Gamma^{\alpha}_{\mu\nu}\Gamma^{\sigma}_{\alpha\sigma} - g^{\mu\sigma}\Gamma^{\alpha}_{\mu\nu}\Gamma^{\nu}_{\sigma\alpha}\right)\sqrt{-g} \\
&= 2g^{\mu\nu}\left(\Gamma^{\alpha}_{\mu\nu}\Gamma^{\beta}_{\alpha\beta} - \Gamma^{\alpha}_{\mu\beta}\Gamma^{\beta}_{\alpha\nu}\right)\sqrt{-g} = 2L\sqrt{-g}.
\end{aligned}
\quad \textbf{(5.12)}
$$

With these terms,

$$R = R^* - L = 2L - L = L, \quad \textbf{(5.13)}$$

the action (5.1) takes the form

$$I = \int \mathcal{L}\mathrm{d}^4x = \int \mathrm{d}x^0 \int \mathcal{L}\mathrm{d}x^1\mathrm{d}x^2\mathrm{d}x^3, \quad \textbf{(5.14)}$$

as an integral of the Lagrangian density

$$\mathcal{L} = L\sqrt{-g} = g^{\mu\nu}\left(\Gamma^{\beta}_{\mu\nu}\Gamma^{\alpha}_{\beta\alpha} - \Gamma^{\beta}_{\mu\alpha}\Gamma^{\alpha}_{\beta\nu}\right)\sqrt{-g}. \quad \textbf{(5.15)}$$

According to the least action principle, we consider a null variation of this action with the variations of the metric elements, and their derivatives with the coordinates:

$$\delta I = \int \delta\mathcal{L}\mathrm{d}^4x = \int\left(\frac{\partial\mathcal{L}}{\partial g_{\alpha\beta}}\delta g_{\alpha\beta} + \frac{\partial\mathcal{L}}{\partial g_{\alpha\beta,\nu}}\delta g_{\alpha\beta,\nu}\right)\mathrm{d}^4x = \int\left[\frac{\partial\mathcal{L}}{\partial g_{\alpha\beta}} - \left(\frac{\partial\mathcal{L}}{\partial g_{\alpha\beta,\nu}}\right)_{,\nu}\right]\delta g_{\alpha\beta}\mathrm{d}^4x = 0, \quad \textbf{(5.16)}$$

which leads to the Lagrange equations

$$\frac{\partial\mathcal{L}}{\partial g_{\alpha\beta}} - \left(\frac{\partial\mathcal{L}}{\partial g_{\alpha\beta,\nu}}\right)_{,\nu} = 0. \quad \textbf{(5.17)}$$

To obtain the two partial derivatives of the Lagrangian density (5.15) with the metric elements and their derivatives with the coordinates, we consider the variation of this density as a function of these elements. With (5.9), the variation of the first term,

$$\delta\left(\Gamma^{\beta}_{\mu\nu}\Gamma^{\alpha}_{\beta\alpha}g^{\mu\nu}\sqrt{-g}\right)=\Gamma^{\beta}_{\mu\nu}\delta\left(\Gamma^{\alpha}_{\beta\alpha}g^{\mu\nu}\sqrt{-g}\right)+\Gamma^{\alpha}_{\beta\alpha}g^{\mu\nu}\sqrt{-g}\delta\Gamma^{\beta}_{\mu\nu}$$
$$=\Gamma^{\beta}_{\mu\nu}\delta\left(g^{\mu\nu}\sqrt{-g}_{,\beta}\right)+\Gamma^{\alpha}_{\beta\alpha}\left[\delta\left(\underline{\Gamma^{\beta}_{\mu\nu}}g^{\mu\nu}\sqrt{-g}\right)-\Gamma^{\beta}_{\mu\nu}\delta\left(g^{\mu\nu}\sqrt{-g}\right)\right], \quad \textbf{(5.18)}$$

takes a form with two variations of the metric elements, and only one variation of a Christoffel symbol. To obtain the variation of this term we calculate the derivative of its factor,

$$\left(g^{\mu\nu}\sqrt{-g}\right)_{,\alpha}=g^{\mu\nu}{}_{,\alpha}\sqrt{-g}+g^{\mu\nu}\sqrt{-g}_{,\alpha}=-g^{\mu\rho}g^{\nu\sigma}g_{\rho\sigma,\alpha}\sqrt{-g}+g^{\mu\nu}\Gamma^{\sigma}_{\alpha\sigma}\sqrt{-g}$$
$$=-g^{\mu\rho}g^{\nu\sigma}\left(\Gamma_{\rho\sigma\alpha}+\Gamma_{\sigma\rho\alpha}\right)\sqrt{-g}+g^{\mu\nu}\Gamma^{\sigma}_{\alpha\sigma}\sqrt{-g}$$
$$=\left(-g^{\nu\sigma}\Gamma^{\mu}_{\sigma\alpha}-g^{\mu\rho}\Gamma^{\nu}_{\rho\alpha}+g^{\mu\nu}\Gamma^{\sigma}_{\alpha\sigma}\right)\sqrt{-g}.$$

By a contraction $\alpha=\nu$, we obtain the expression

$$\left(g^{\mu\nu}\sqrt{-g}\right)_{,\nu}=\left(-g^{\nu\sigma}\Gamma^{\mu}_{\sigma\nu}-\cancel{g^{\mu\rho}\Gamma^{\nu}_{\rho\nu}}+\cancel{g^{\mu\nu}\Gamma^{\sigma}_{\nu\sigma}}\right)\sqrt{-g}=-g^{\nu\sigma}\Gamma^{\mu}_{\sigma\nu}\sqrt{-g},$$

which, for an interchange of the indices $\mu\leftrightarrow\sigma$, and a notation $\beta=\sigma$, we the second term of (5.18),

$$g^{\mu\nu}\Gamma^{\beta}_{\mu\nu}\sqrt{-g}=-\left(g^{\beta\nu}\sqrt{-g}\right)_{,\nu}.$$

With this expression, the variation term (5.18) of the Lagrangian density (5.15) takes a form depending only on variations of the metric elements:

$$\delta\left(\Gamma^{\beta}_{\mu\nu}\Gamma^{\alpha}_{\beta\alpha}g^{\mu\nu}\sqrt{-g}\right)=\Gamma^{\beta}_{\mu\nu}\delta\left(g^{\mu\nu}\sqrt{-g}_{,\beta}\right)-\Gamma^{\alpha}_{\beta\alpha}\delta\left(g^{\beta\nu}\sqrt{-g}\right)_{,\nu}-\Gamma^{\alpha}_{\beta\alpha}\Gamma^{\beta}_{\mu\nu}\delta\left(g^{\mu\nu}\sqrt{-g}\right). \quad \textbf{(5.19)}$$

Similarly, we calculate the variation of the second term of the Lagrangian density (5.15). With an index interchange $\alpha\leftrightarrow\beta$, we obtain

$$\begin{aligned}\delta\left(\Gamma^{\beta}_{\mu\alpha}\Gamma^{\alpha}_{\beta\nu}g^{\mu\nu}\sqrt{-g}\right)&=\underline{\delta\left(\Gamma^{\beta}_{\mu\alpha}\right)\Gamma^{\alpha}_{\beta\nu}g^{\mu\nu}\sqrt{-g}}+\underline{\Gamma^{\beta}_{\mu\alpha}\delta\left(\Gamma^{\alpha}_{\beta\nu}\right)g^{\mu\nu}\sqrt{-g}}+\Gamma^{\beta}_{\mu\alpha}\Gamma^{\alpha}_{\beta\nu}\delta\left(g^{\mu\nu}\sqrt{-g}\right)\\&=2\delta\left(\Gamma^{\beta}_{\mu\alpha}\right)\Gamma^{\alpha}_{\beta\nu}g^{\mu\nu}\sqrt{-g}+\Gamma^{\beta}_{\mu\alpha}\Gamma^{\alpha}_{\beta\nu}\delta\left(g^{\mu\nu}\sqrt{-g}\right)\\&=2\delta\left(\Gamma^{\beta}_{\mu\alpha}g^{\mu\nu}\sqrt{-g}\right)\Gamma^{\alpha}_{\beta\nu}-2\Gamma^{\beta}_{\mu\alpha}\Gamma^{\alpha}_{\beta\nu}\delta\left(g^{\mu\nu}\sqrt{-g}\right)+\Gamma^{\beta}_{\mu\alpha}\Gamma^{\alpha}_{\beta\nu}\delta\left(g^{\mu\nu}\sqrt{-g}\right)\\&=2\delta\left(\Gamma^{\beta}_{\mu\alpha}g^{\mu\nu}\sqrt{-g}\right)\Gamma^{\alpha}_{\beta\nu}-\Gamma^{\beta}_{\mu\alpha}\Gamma^{\alpha}_{\beta\nu}\delta\left(g^{\mu\nu}\sqrt{-g}\right)\\&=\delta\left(g^{\mu\nu}\Gamma^{\beta}_{\mu\alpha}\sqrt{-g}+g^{\mu\nu}\Gamma^{\beta}_{\mu\alpha}\sqrt{-g}\right)\Gamma^{\alpha}_{\beta\nu}-\Gamma^{\beta}_{\mu\alpha}\Gamma^{\alpha}_{\beta\nu}\delta\left(g^{\mu\nu}\sqrt{-g}\right).\end{aligned}$$

By an index interchange $\beta\leftrightarrow\nu$, and the symmetry of the Christoffel symbol $\Gamma^{\alpha}_{\beta\nu}=\Gamma^{\alpha}_{\beta\nu}$, this expression becomes

$$\delta\left(\Gamma^{\beta}_{\mu\alpha}\Gamma^{\alpha}_{\beta\nu}g^{\mu\nu}\sqrt{-g}\right)=\delta\left[\left(g^{\mu\nu}\Gamma^{\beta}_{\mu\alpha}+g^{\mu\beta}\Gamma^{\nu}_{\mu\alpha}\right)\sqrt{-g}\right]\Gamma^{\alpha}_{\beta\nu}-\Gamma^{\beta}_{\mu\alpha}\Gamma^{\alpha}_{\beta\nu}\delta\left(g^{\mu\nu}\sqrt{-g}\right).\quad\textbf{(5.20)}$$

At the same time, from the derivative of the contravariant metric tensor,

$$g^{\mu\nu}{}_{,\alpha}=-g^{\mu\rho}g^{\nu\sigma}g_{\rho\sigma,\alpha}=-g^{\mu\rho}g^{\nu\sigma}\left(\Gamma_{\rho\sigma\alpha}+\Gamma_{\sigma\rho\alpha}\right)=-g^{\nu\sigma}\Gamma^{\mu}_{\sigma\alpha}-g^{\mu\rho}\Gamma^{\nu}_{\rho\alpha},$$

with a change of indices, we obtain the variation of the first term of the variation (5.20):

$$g^{\mu\nu}\Gamma^{\beta}_{\mu\alpha}+g^{\mu\beta}\Gamma^{\nu}_{\mu\alpha}=g^{\nu\beta}{}_{,\alpha}.$$

Thus, this variation becomes

$$\delta\left(\Gamma^{\beta}_{\mu\alpha}\Gamma^{\alpha}_{\beta\nu}g^{\mu\nu}\sqrt{-g}\right)=-\delta\left(g^{\nu\beta}{}_{,\alpha}\sqrt{-g}\right)\Gamma^{\alpha}_{\beta\nu}-\Gamma^{\beta}_{\mu\alpha}\Gamma^{\alpha}_{\beta\nu}\delta\left(g^{\mu\nu}\sqrt{-g}\right).\quad\textbf{(5.21)}$$

With (5.19) and (5.21), we obtain the variation (5.15) of the Lagrangian density:

$$\begin{aligned}\delta\mathcal{L} &= \delta\left(\Gamma^{\beta}_{\mu\nu}\Gamma^{\alpha}_{\beta\alpha}g^{\mu\nu}\sqrt{-g}\right)-\delta\left(\Gamma^{\beta}_{\mu\alpha}\Gamma^{\alpha}_{\beta\nu}g^{\mu\nu}\sqrt{-g}\right)\\ &= \Gamma^{\beta}_{\mu\nu}\delta\left(g^{\mu\nu}\sqrt{-g}_{,\beta}\right)-\Gamma^{\alpha}_{\beta\alpha}\delta\left(g^{\beta\nu}\sqrt{-g}\right)_{,\nu}-\Gamma^{\alpha}_{\beta\alpha}\Gamma^{\beta}_{\mu\nu}\delta\left(g^{\mu\nu}\sqrt{-g}\right)\\ &\quad+\delta\left(g^{\nu\beta}{}_{,\alpha}\sqrt{-g}\right)\Gamma^{\alpha}_{\beta\nu}+\Gamma^{\beta}_{\mu\alpha}\Gamma^{\alpha}_{\beta\nu}\delta\left(g^{\mu\nu}\sqrt{-g}\right)\\ &= \underline{\Gamma^{\alpha}_{\mu\nu}\delta\left(g^{\mu\nu}\sqrt{-g}_{,\alpha}\right)}-\Gamma^{\alpha}_{\beta\alpha}\delta\left(g^{\beta\nu}\sqrt{-g}\right)_{,\nu}-\Gamma^{\alpha}_{\beta\alpha}\Gamma^{\beta}_{\mu\nu}\delta\left(g^{\mu\nu}\sqrt{-g}\right)\\ &\quad\underline{+\Gamma^{\alpha}_{\mu\nu}\delta\left(g^{\nu\mu}{}_{,\alpha}\sqrt{-g}\right)}+\Gamma^{\beta}_{\mu\alpha}\Gamma^{\alpha}_{\beta\nu}\delta\left(g^{\mu\nu}\sqrt{-g}\right).\end{aligned}$$

Taking the first and the fourth terms for the derivative of a product, we obtain this variation as a function of the metric element variations:

$$\begin{aligned}\delta\mathcal{L} &= \Gamma^{\alpha}_{\mu\nu}\delta\left(g^{\mu\nu}\sqrt{-g}\right)_{,\alpha}-\Gamma^{\alpha}_{\beta\alpha}\delta\left(g^{\beta\nu}\sqrt{-g}\right)_{,\nu}+\left(\Gamma^{\beta}_{\mu\alpha}\Gamma^{\alpha}_{\beta\nu}-\Gamma^{\alpha}_{\beta\alpha}\Gamma^{\beta}_{\mu\nu}\right)\delta\left(g^{\mu\nu}\sqrt{-g}\right)\\ &= \Gamma^{\alpha}_{\mu\nu}\delta\left(g^{\mu\nu}\sqrt{-g}\right)_{,\alpha}-\Gamma^{\beta}_{\mu\beta}\delta\left(g^{\mu\nu}\sqrt{-g}\right)_{,\nu}+\left(\Gamma^{\beta}_{\mu\alpha}\Gamma^{\alpha}_{\beta\nu}-\Gamma^{\beta}_{\mu\nu}\Gamma^{\alpha}_{\beta\alpha}\right)\delta\left(g^{\mu\nu}\sqrt{-g}\right)\\ &= \Gamma^{\alpha}_{\mu\nu}\delta\left(g^{\mu\nu}\sqrt{-g}\right)_{,\alpha}-g^{\alpha}_{\nu}\Gamma^{\beta}_{\mu\beta}\delta\left(g^{\mu\nu}\sqrt{-g}\right)_{,\alpha}+\left(\Gamma^{\alpha}_{\mu\beta}\Gamma^{\beta}_{\alpha\nu}-\Gamma^{\alpha}_{\mu\nu}\Gamma^{\beta}_{\alpha\beta}\right)\delta\left(g^{\mu\nu}\sqrt{-g}\right)\\ &= \left(\Gamma^{\alpha}_{\mu\nu}-g^{\alpha}_{\nu}\Gamma^{\beta}_{\mu\beta}\right)\delta\left(g^{\mu\nu}\sqrt{-g}\right)_{,\alpha}+\left(\Gamma^{\alpha}_{\mu\beta}\Gamma^{\beta}_{\alpha\nu}-\Gamma^{\alpha}_{\mu\nu}\Gamma^{\beta}_{\alpha\beta}\right)\delta\left(g^{\mu\nu}\sqrt{-g}\right)\\ &= \frac{\partial\mathcal{L}}{\partial\left(g^{\mu\nu}\sqrt{-g}\right)}\delta\left(g^{\mu\nu}\sqrt{-g}\right)+\frac{\partial\mathcal{L}}{\partial\left(g^{\mu\nu}\sqrt{-g}\right)_{,\alpha}}\delta\left(g^{\mu\nu}\sqrt{-g}\right)_{,\alpha}\end{aligned} \quad (5.22)$$

which leads to the terms of the Lagrange equation (5.17)

$$\begin{aligned}\frac{\partial\mathcal{L}}{\partial\left(g^{\mu\nu}\sqrt{-g}\right)} &= \Gamma^{\alpha}_{\mu\beta}\Gamma^{\beta}_{\alpha\nu}-\Gamma^{\alpha}_{\mu\nu}\Gamma^{\beta}_{\alpha\beta}\\ \frac{\partial\mathcal{L}}{\partial\left(g^{\mu\nu}\sqrt{-g}\right)_{,\alpha}} &= \Gamma^{\alpha}_{\mu\nu}-g^{\alpha}_{\nu}\Gamma^{\beta}_{\mu\beta}.\end{aligned} \quad (5.23)$$

With the Lagrangian (5.15) and the second expression (5.23), we define the pseudo-energy tensor

$$
\begin{aligned}
t_\sigma^\rho &= \frac{\partial \mathcal{L}}{\partial \left(g^{\mu\nu}\sqrt{-g}\right)_{,\rho}} \left(g^{\mu\nu}\sqrt{-g}\right)_{,\sigma} - g_\sigma^\rho \mathcal{L} \\
&= \left(\Gamma_{\mu\nu}^{\rho} - g_\nu^\rho \Gamma_{\mu\alpha}^{\alpha}\right)\left(g^{\mu\nu}\sqrt{-g}\right)_{,\sigma} - g_\sigma^\rho \left(\Gamma_{\mu\nu}^{\beta}\Gamma_{\beta\alpha}^{\alpha} - \Gamma_{\mu\alpha}^{\beta}\Gamma_{\beta\nu}^{\alpha}\right) g^{\mu\nu}\sqrt{-g}
\end{aligned}
\tag{5.24}
$$

At the same time, the Lagrangian variation (5.22),

$$
\begin{aligned}
\delta\mathcal{L} &= \Gamma_{\mu\nu}^{\alpha}\delta\left(g^{\mu\nu}\sqrt{-g}\right)_{,\alpha} - \Gamma_{\mu\beta}^{\beta}\delta\left(g^{\mu\nu}\sqrt{-g}\right)_{,\nu} + \left(\Gamma_{\mu\alpha}^{\beta}\Gamma_{\beta\nu}^{\alpha} - \Gamma_{\mu\nu}^{\beta}\Gamma_{\beta\alpha}^{\alpha}\right)\delta\left(g^{\mu\nu}\sqrt{-g}\right) \\
&= \left[\Gamma_{\mu\nu}^{\alpha}\delta\left(g^{\mu\nu}\sqrt{-g}\right)\right]_{,\alpha} \underline{\underline{- \Gamma_{\mu\nu,\alpha}^{\alpha}\delta\left(g^{\mu\nu}\sqrt{-g}\right)}} - \left[\Gamma_{\mu\beta}^{\beta}\delta\left(g^{\mu\nu}\sqrt{-g}\right)\right]_{,\nu} \underline{+ \Gamma_{\mu\beta,\nu}^{\beta}\delta\left(g^{\mu\nu}\sqrt{-g}\right)} \\
&\quad \underline{\underline{\underline{+ \left(\Gamma_{\mu\alpha}^{\beta}\Gamma_{\beta\nu}^{\alpha} - \Gamma_{\mu\nu}^{\beta}\Gamma_{\beta\alpha}^{\alpha}\right)\delta\left(g^{\mu\nu}\sqrt{-g}\right)}}}
\end{aligned}
$$

with the curvature Ricci tensor (5.3) multiplied with the metric tensor variation

$$
R_{\mu\nu}\delta\left(g^{\mu\nu}\sqrt{-g}\right) = \underline{\Gamma_{\mu\alpha,\nu}^{\alpha}\delta\left(g^{\mu\nu}\sqrt{-g}\right)} \underline{\underline{- \Gamma_{\mu\nu,\alpha}^{\alpha}\delta\left(g^{\mu\nu}\sqrt{-g}\right)}} \underline{\underline{\underline{+ \left(\Gamma_{\mu\alpha}^{\beta}\Gamma_{\beta\nu}^{\alpha} - \Gamma_{\mu\nu}^{\beta}\Gamma_{\beta\alpha}^{\alpha}\right)\delta\left(g^{\mu\nu}\sqrt{-g}\right)}}}
$$

takes the form

$$
\delta\mathcal{L} = \left[\Gamma_{\mu\nu}^{\alpha}\delta\left(g^{\mu\nu}\sqrt{-g}\right)\right]_{,\alpha} - \left[\Gamma_{\mu\beta}^{\beta}\delta\left(g^{\mu\nu}\sqrt{-g}\right)\right]_{,\nu} + R_{\mu\nu}\delta\left(g^{\mu\nu}\sqrt{-g}\right)
$$

including perfect differentials which have no contribution to the action integral (5.14). We obtain

$$
\delta I = \int \delta\mathcal{L}\,\mathrm{d}^4x = \int R_{\mu\nu}\delta\left(g^{\mu\nu}\sqrt{-g}\right)\mathrm{d}^4x = 0, \tag{5.25}
$$

which leads to a null Ricci tensor:

$$
R^{\mu\nu} = 0. \tag{5.26}
$$

With the variations obtained from (3.46) and (3.66),

$$\delta g^{\mu\nu} = -g^{\mu\alpha} g^{\nu\beta} \delta g_{\alpha\beta}$$
$$\delta\sqrt{-g} = \frac{1}{2}\sqrt{-g} g^{\alpha\beta} \delta g_{\alpha\beta}, \tag{5.27}$$

we obtain

$$\begin{aligned}\delta\left(g^{\mu\nu}\sqrt{-g}\right) &= \sqrt{-g}\delta g^{\mu\nu} + g^{\mu\nu}\delta\sqrt{-g} = -\sqrt{-g} g^{\mu\alpha} g^{\nu\beta} \delta g_{\alpha\beta} + g^{\mu\nu}\frac{1}{2}\sqrt{-g} g^{\alpha\beta} \delta g_{\alpha\beta} \\ &= -\left(g^{\mu\alpha} g^{\nu\beta} - \frac{1}{2} g^{\mu\nu} g^{\alpha\beta}\right)\sqrt{-g}\delta g_{\alpha\beta}.\end{aligned} \tag{5.28}$$

With this expression, the action variation (5.25) takes the form

$$\begin{aligned}\mathrm{d}I &= -\int R_{\mu\nu}\left(g^{\mu\alpha} g^{\nu\beta} - \frac{1}{2} g^{\mu\nu} g^{\alpha\beta}\right)\sqrt{-g}\delta g_{\alpha\beta}\mathrm{d}^4x \\ &= -\int\left(R^{\alpha\beta} - \frac{1}{2} R g^{\alpha\beta}\right)\sqrt{-g}\delta g_{\alpha\beta}\mathrm{d}^4x = 0,\end{aligned} \tag{5.29}$$

which leads to Einstein's equation of gravity in vacuum (3.123).

5.2. MATTER ACTION

We consider the matter action as the integral of the matter density over the four-dimensional physical space,

$$I_m = -\int \rho\sqrt{-g}\mathrm{d}^4x. \tag{5.30}$$

To describe the matter dynamics, we consider the momentum four-vector density

$$p^\mu = \rho v^\mu \sqrt{-g}, \tag{5.31}$$

with the fundamental equation (1.48),

$$v_\mu v^\mu = 1. \tag{5.32}$$

With this expression, we obtain the scalar density function

$$\rho\sqrt{-g}=\left(\rho v^{\mu}\sqrt{-g}\rho v_{\mu}\sqrt{-g}\right)^{1/2}=\left(p^{\mu}p_{\mu}\right)^{1/2}, \qquad \textbf{(5.33)}$$

as the action integral (5.30) is

$$I_m=-\int\left(p^{\mu}p_{\mu}\right)^{1/2}\mathrm{d}^4x. \qquad \textbf{(5.34)}$$

With the variation of the density function (5.33),

$$\begin{aligned}\delta\left(p^{\mu}p_{\mu}\right)^{1/2}&=\frac{1}{2}\left(p^{\lambda}p_{\lambda}\right)^{-1/2}\left(p^{\mu}p^{\nu}\delta g_{\mu\nu}+2p_{\mu}\delta p^{\mu}\right)\\&=\frac{1}{2}\left(\rho\sqrt{-g}\right)^{-1}\left(\rho^{2}v^{\mu}v^{\nu}\sqrt{-g}^{2}\delta g_{\mu\nu}+2\rho v_{\mu}\sqrt{-g}\,\delta p^{\mu}\right)\\&=\frac{1}{2}\rho v^{\mu}v^{\nu}\sqrt{-g}\delta g_{\mu\nu}+v_{\mu}\delta p^{\mu},\end{aligned}$$

the we obtain the variation of the total action I_g+I_m,

$$\delta\left(I_g+I_m\right)=-\int\left[\frac{1}{16\pi}\left(R^{\mu\nu}-\frac{1}{2}g^{\mu\nu}R\right)+\frac{1}{2}\rho v^{\mu}v^{\nu}\right]\sqrt{\ }\delta g_{\mu\nu}\mathrm{d}^4x-\int v_{\mu}\delta p^{\mu}\mathrm{d}^4x, \qquad \textbf{(5.35)}$$

with a coefficient of the gravitational action (5.29), in agreement with Einstein's gravitation law (3.198):

$$I_g=\frac{1}{16\pi}I.$$

We consider the momentum variation δp^{μ} with the coordinate variations δx^{ν}, of a form satisfying the principle of inertia, $\delta p^{\mu}=0$ for a variation in the direction of this momentum, $\delta x^{\nu}=\delta x^{\mu}$, $\nu=\mu$:

$$\delta p^{\mu}=p^{\nu}{}_{,\nu}\delta x^{\mu}-p^{\mu}{}_{,\nu}\delta x^{\nu}=\left(p^{\nu}\delta x^{\mu}-p^{\mu}\delta x^{\nu}\right)_{,\nu}. \qquad \textbf{(5.36)}$$

With (5.36), the last term of the action variation (5.35) is

$$
\begin{aligned}
-\int v_{\mu}\delta p^{\mu}\mathrm{d}^4x &= -\int v_{\mu}\left(p^{\nu}\delta x^{\mu}-p^{\mu}\delta x^{\nu}\right)_{,\nu}\mathrm{d}^4x \\
&= \int v_{\mu,\nu}\left(p^{\nu}\delta x^{\mu}-p^{\mu}\delta x^{\nu}\right)\mathrm{d}^4x \\
&= \int\left(v_{\mu,\nu}p^{\nu}\delta x^{\mu}-v_{\nu,\mu}p^{\nu}\delta x^{\mu}\right)\mathrm{d}^4x \\
&= \int\left(v_{\mu,\nu}-v_{\nu,\mu}\right)p^{\nu}\delta x^{\mu}\mathrm{d}^4x
\end{aligned}
$$

Considering the ordinary differentials as functions of the covariant differentials,

$$
v_{\mu:\nu}=v_{\mu,\nu}-\underline{\Gamma^{\alpha}_{\mu\nu}v_{\alpha}}, \qquad v_{\nu:\mu}=v_{\nu,\mu}-\underline{\Gamma^{\alpha}_{\nu\mu}v_{\alpha}},
$$

we obtain

$$
-\int v_{\mu}\delta p^{\mu}\mathrm{d}^4x=\int\left(v_{\mu:\nu}-v_{\nu:\mu}\right)\rho v^{\nu}\delta x^{\mu}\sqrt{-g}\mathrm{d}^4x. \tag{5.37}
$$

From the covariant derivative of the fundamental equation (1.48),

$$
v_{\nu:\mu}v^{\nu}+v_{\nu}v^{\nu}{}_{:\mu}=2v_{\nu:\mu}v^{\nu}=0,
$$

we obtain that the velocity in the proper time is perpendicular to any of its covariant derivatives, as the total variation (5.35) takes a form depending the metric tensor and coordinate variations:

$$
\delta\left(I_g+I_m\right)=-\int\left[\frac{1}{16\pi}\left(R^{\mu\nu}-\frac{1}{2}g^{\mu\nu}R\right)+\frac{1}{2}\rho v^{\mu}v^{\nu}\right]\sqrt{-g}\delta g_{\mu\nu}\mathrm{d}^4x+\int\rho v_{\mu:\nu}v^{\nu}\delta x^{\mu}\sqrt{-g}\mathrm{d}^4x. \tag{5.38}
$$

From a null action variation (5.38), we obtain the geodesic equation of the form (3.91) of a null covariant acceleration:

$$
v_{\mu:\nu}v^{\nu}=0. \tag{5.39}
$$

5.3. ELECTROMAGNETIC FIELD ACTION

We consider the Maxwell equations (4.42),

$$\begin{aligned}
&\frac{1}{c}\frac{\partial}{\partial t}\vec{E} = \nabla \times \vec{H} - 4\pi \tilde{\tilde{\vec{j}}} \\
&\frac{1}{c}\frac{\partial}{\partial t}\vec{H} = -\nabla \times \vec{E} \\
&\nabla \vec{E} = 4\pi\tilde{\rho} \\
&\nabla \vec{H} = 0,
\end{aligned} \tag{5.40}$$

for the electric field (4.8),

$$\vec{E} = -\nabla A^0 - \frac{\partial}{\partial t}\vec{A}, \tag{5.41}$$

and the magnetic field (4.9) with (4.23) and (4.37),

$$\vec{H} = c\nabla \times \vec{A}, \tag{5.42}$$

as functions of the electromagnetic potential

$$A^0 = U, \quad c\vec{A} = \left(A^1, A^2, A^3\right), \tag{5.43}$$

with the gauge condition (4.13),

$$\nabla \vec{A} = 0. \tag{5.44}$$

For a flat space,

$$g_{00} = 1, \quad g_{11} = -1, \quad g_{22} = -1, \quad g_3 = -1, \tag{5.45}$$

we obtain the covariant electromagnetic potential:

$$A_0 = A^0, \quad A_1 = -A^1, \quad A_2 = -A^2, \quad A_3 = -A^3. \tag{5.46}$$

With these components, for the electromagnetic field we obtain a tensorial representation:

$$F_{\mu\nu} = A_{\mu,\nu} - A_{\nu,\mu} \,. \tag{5.47}$$

From (5.41) we obtain the electric field,

$$\begin{aligned} E^1 &= -\frac{1}{c}\frac{\partial A^1}{\partial t} - \frac{\partial A^0}{\partial x^1} = -\frac{\partial A^1}{\partial x^0} - \frac{\partial A^0}{\partial x^1} = \frac{\partial A_1}{\partial x^0} - \frac{\partial A_0}{\partial x^1} = F_{10} = -F^{10} \\ E^2 &= -\frac{1}{c}\frac{\partial A^2}{\partial t} - \frac{\partial A^0}{\partial x^2} = \frac{\partial A_2}{\partial x^0} - \frac{\partial A_0}{\partial x^2} = F_{20} = -F^{20} \\ E^3 &= -\frac{1}{c}\frac{\partial A^3}{\partial t} - \frac{\partial A^0}{\partial x^3} = \frac{\partial A_3}{\partial x^0} - \frac{\partial A_0}{\partial x^3} = F_{30} = -F^{30}, \end{aligned} \tag{5.48}$$

and from (5.42) with (5.43) we obtain the magnetic field:

$$\begin{aligned} H^1 &= \frac{\partial A^3}{\partial x^2} - \frac{\partial A^2}{\partial x^3} = \frac{\partial A_2}{\partial x^3} - \frac{\partial A_3}{\partial x^2} = F_{23} = F^{23} \\ H^2 &= \frac{\partial A^1}{\partial x^3} - \frac{\partial A^3}{\partial x^1} = \frac{\partial A_3}{\partial x^1} - \frac{\partial A_1}{\partial x^3} = F_{31} = F^{31} \\ H^3 &= \frac{\partial A^2}{\partial x^1} - \frac{\partial A^1}{\partial x^2} = \frac{\partial A_1}{\partial x^2} - \frac{\partial A_2}{\partial x^1} = F_{12} = F^{12}. \end{aligned} \tag{5.49}$$

With these elements, we obtain the electromagnetic field tensor:

$$F^{\mu\nu} = \begin{pmatrix} 0 & E^1 & E^2 & E^3 \\ -E^1 & 0 & H^3 & -H^2 \\ -E^2 & -H^3 & 0 & H^1 \\ -E^3 & H^2 & -H^1 & 0 \end{pmatrix}. \tag{5.50}$$

From the divergence of the four line vectors of this matrix, we obtain equations

$$F^{0\nu}{}_{,\nu} = F^{0m}{}_{,m} = E^{m}{}_{,m} = E^{1}{}_{,1} + E^{2}{}_{,2} + E^{3}{}_{,3} = 4\pi\tilde{\rho}$$
$$F^{1\nu}{}_{,\nu} = F^{10}{}_{,0} + F^{12}{}_{,2} + F^{13}{}_{,3} = -E^{1}{}_{,0} + H^{3}{}_{,2} - H^{2}{}_{,3} = 4\pi\tilde{j}^{1}$$
$$F^{2\nu}{}_{,\nu} = F^{20}{}_{,0} + F^{21}{}_{,1} + F^{23}{}_{,3} = -E^{2}{}_{,0} - H^{3}{}_{,1} + H^{1}{}_{,3} = 4\pi\tilde{j}^{2}$$
$$F^{3\nu}{}_{,\nu} = F^{30}{}_{,0} + F^{31}{}_{,1} + F^{32}{}_{,2} = -E^{3}{}_{,0} + H^{2}{}_{,1} - H^{1}{}_{,2} = 4\pi\tilde{j}^{3}$$

which are of the form of the third Maxwell equation (5.40),

$$\nabla\vec{E} = 4\pi\tilde{\rho}\,,$$

and of the first equation (5.40),

$$\frac{1}{c}\frac{\partial\vec{E}}{\partial t} = \nabla\times\vec{H} - 4\pi\tilde{\vec{j}}\,.$$

For the tensor elements (5.47), by cyclic permutations we obtain the coordinate derivatives:

$$F_{\mu\nu,\sigma} = \underline{A_{\mu,\nu\sigma}} - \underline{\underline{A_{\nu,\mu\sigma}}}$$
$$F_{\nu\sigma,\mu} = \underline{\underline{A_{\nu,\sigma\mu}}} - A_{\sigma,\nu\mu}$$
$$F_{\sigma\mu,\nu} = A_{\sigma,\mu\nu} - \underline{A_{\mu,\sigma\nu}}\,,$$

which lead to the electromagnetic field tensorial equation

$$F_{\mu\nu,\sigma} + F_{\nu\sigma,\mu} + F_{\sigma\mu,\nu} = 0\,, \qquad \textbf{(5.51)}$$

as a function of the elements of the covariant tensor, which, from (5.50) with (5.48) and (5.49) gets the form:

$$F_{\mu\nu} = \begin{pmatrix} 0 & -E^1 & -E^2 & -E^3 \\ E^1 & 0 & H^3 & -H^2 \\ E^2 & -H^3 & 0 & H^1 \\ E^3 & H^2 & -H^1 & 0 \end{pmatrix}. \quad \textbf{(5.52)}$$

With the elements of this tensor, from the electromagnetic field tensorial equation (5.51) we obtain equations

$$F_{01,2} + F_{12,0} + F_{20,1} = -E^1{}_{,2} + H^3{}_{,0} + E^2{}_{,1} = 0, \quad \frac{\partial H^3}{\partial x^0} = -\frac{\partial E^2}{\partial x^1} + \frac{\partial E^1}{\partial x^2}$$

$$F_{02,3} + F_{23,0} + F_{30,2} = -E^2{}_{,3} + H^1{}_{,0} + E^3{}_{,2} = 0, \quad \frac{\partial H^1}{\partial x^0} = -\frac{\partial E^3}{\partial x^2} + \frac{\partial E^2}{\partial x^3}$$

$$F_{03,1} + F_{31,0} + F_{10,3} = -E^3{}_{,1} + H^2{}_{,0} + E^1{}_{,3} = 0, \quad \frac{\partial H^2}{\partial x^0} = -\frac{\partial E^1}{\partial x^3} + \frac{\partial E^3}{\partial x^1},$$

with the vector form of the second Maxwell equation (5.40),

$$\frac{1}{c}\frac{\partial \vec{H}}{\partial t} = -\nabla \times \vec{E},$$

and

$$F_{12,3} + F_{23,1} + F_{31,2} = H^3{}_{,3} + H^1{}_{,1} + H^2{}_{,2} = 0,$$

which is of the vector form of the fourth Maxwell equation (5.40):

$$\nabla \vec{H} = 0.$$

Thus, with the electromagnetic field tensor (5.50), or (5.52), in a flat space, we obtained the Maxwell equations (5.40).

We consider an electromagnetic field (5.47) of the electromagnetic potential (5.46) in the general form of a curved space, which, according to (3.88) being a difference of two covariant derivatives,

$$F_{\mu\nu} = A_{\mu,\nu} - A_{\nu,\mu} = A_{\mu:\nu} - A_{\nu:\mu}, \tag{5.53}$$

is a tensor, and define the electromagnetic field action

$$I = \int F_{\mu\nu} F^{\nu\mu} \sqrt{-g}\,\mathrm{d}^4 x. \tag{5.54}$$

From the variation of this action with the metric tensor, we obtain

$$\begin{aligned}
\delta\left(F_{\mu\nu}F^{\nu\mu}\sqrt{-g}\right) &= F_{\mu\nu}F^{\nu\mu}\delta\sqrt{-g} + \delta\left(F_{\mu\nu}F^{\nu\mu}\right)\sqrt{-g} \\
&= F_{\mu\nu}F^{\nu\mu}\delta\sqrt{-g} + \delta\left(F_{\mu\nu}\, F_{\beta\alpha}g^{\beta\nu}g^{\alpha\mu}\right)\sqrt{-g} \\
&= F_{\mu\nu}F^{\nu\mu}\delta\sqrt{-g} + F_{\mu\nu}\, F_{\beta\alpha}\sqrt{-g}\,\delta\left(g^{\beta\nu}g^{\alpha\mu}\right) \\
&= F_{\mu\nu}F^{\nu\mu}\frac{1}{2}\sqrt{-g}g^{\rho\sigma}\delta g_{\rho\sigma} + F_{\mu\nu}\, F_{\alpha\beta}\sqrt{-g}\left(g^{\beta\nu}\delta g^{\alpha\mu} + g^{\alpha\mu}\delta g^{\beta\nu}\right).
\end{aligned} \tag{5.55}$$

The last term of this expression is of the form

$$\begin{aligned}
F_{\mu\nu}\, F_{\beta\alpha}\left(g^{\nu\beta}\delta g^{\mu\alpha} + g^{\mu\alpha}\delta g^{\nu\beta}\right) &= F_{\mu\nu}\, F_{\beta\alpha}\left(-g^{\nu\beta}g^{\mu\rho}g^{\alpha\sigma}\delta g_{\rho\sigma} - g^{\mu\alpha}g^{\nu\rho}g^{\beta\sigma}\delta g_{\rho\sigma}\right) \\
&= F_{\mu\nu}\, F_{\beta\alpha}\left(-g^{\nu\beta}g^{\mu\rho}g^{\alpha\sigma} - g^{\mu\alpha}g^{\nu\rho}g^{\beta\sigma}\right)\delta g_{\rho\sigma}.
\end{aligned}$$

Since the two field matrices are antisymmetric, by an index interchange of the of the second term, we obtain

$$\begin{aligned}
F_{\mu\nu}\, F_{\beta\alpha}\left(g^{\nu\beta}\delta g^{\mu\alpha} + g^{\mu\alpha}\delta g^{\nu\beta}\right) &= F_{\mu\nu}\, F_{\beta\alpha}\left(-g^{\nu\beta}g^{\mu\rho}g^{\alpha\sigma} - g^{\nu\beta}g^{\mu\rho}g^{\alpha\sigma}\right)\delta g_{\rho\sigma} \\
&= -2F_{\mu\nu}\, F_{\beta\alpha}g^{\mu\rho}g^{\alpha\sigma}g^{\nu\beta}\delta g_{\rho\sigma}.
\end{aligned}$$

With this expression, the variation (5.55) of the Lagrangian density of the action (5.54) is

$$\delta\left(F_{\mu\nu}F^{\nu\mu}\sqrt{-g}\right) = \left(F_{\mu\nu}F^{\nu\mu}\frac{1}{2}g^{\rho\sigma} - 2F^{\rho}{}_{\nu}\, F^{\nu\sigma}\right)\sqrt{-g}\,\delta g_{\rho\sigma}. \tag{5.56}$$

For a physical understanding of this Lagrangian density, we consider the field tensors (5.52),

$$F_{\mu\nu} = \begin{pmatrix} 0 & -E^1 & -E^2 & -E^3 \\ E^1 & 0 & H^3 & -H^2 \\ E^2 & -H^3 & 0 & H^1 \\ E^3 & H^2 & -H^1 & 0 \end{pmatrix}, \tag{5.57}$$

and (5.50) with its antisymmetric counterpart,

$$F^{\mu\nu} = \begin{pmatrix} 0 & E^1 & E^2 & E^3 \\ -E^1 & 0 & H^3 & -H^2 \\ -E^2 & -H^3 & 0 & H^1 \\ -E^3 & H^2 & -H^1 & 0 \end{pmatrix}, \quad F^{\nu\mu} = \begin{pmatrix} 0 & -E^1 & -E^2 & -E^3 \\ E^1 & 0 & -H^3 & H^2 \\ E^2 & H^3 & 0 & -H^1 \\ E^3 & -H^2 & H^1 & 0 \end{pmatrix}. \tag{5.58}$$

and calculate the product

$$\begin{aligned} F_{\mu\nu}F^{\nu\mu} &= \begin{pmatrix} -\vec{E}^2 & -E^2H^3+E^3H^2 & -E^3H^1+E^1H^3 & -E^1H^2+E^2H^1 \\ E^2H^3-E^3H^2 & -E^{1^2}+H^{3^2}+H^{2^2} & -E^1E^2-H^1H^2 & -E^1E^3-H^1H^3 \\ -E^1H^3+E^3H^1 & -E^1E^2-H^1H^2 & -E^{2^2}+H^{3^2}+H^{1^2} & -E^2E^3-H^2H^3 \\ -E^1H^2+E^2H^1 & -E^1E^3-H^1H^3 & -E^2E^3-H^2H^3 & -E^{3^2}+H^{2^2}+H^{1^2} \end{pmatrix} \\ &= -\vec{E}^2 - E^{1^2} + H^{3^2} + H^{2^2} - E^{2^2} + H^{3^2} + H^{1^2} - E^{3^2} + H^{2^2} + H^{1^2} \\ &= -2\left(\vec{E}^2 - \vec{H}^2\right). \end{aligned} \tag{5.59}$$

With this expression, the action integral (5.54) with the metric tensor (5.45) takes the form

$$I = -2\int\left(\vec{E}^2 - \vec{H}^2\right)\mathrm{d}^4x, \tag{5.60}$$

as the variation coefficient (5.56) is

$$F_{\mu\nu}F^{\nu\mu}\frac{1}{2}g^{\rho\sigma}-2F^{\rho}{}_{\nu}F^{\nu\sigma}=-\left(\vec{E}^{2}-\vec{H}^{2}\right)g^{\rho\sigma}-2F^{\rho}{}_{\nu}F^{\nu\sigma}. \qquad \textbf{(5.61)}$$

From (5.57) and (5.58), we obtain,

$$F^{0}{}_{\nu}=g^{0\mu}F_{\mu\nu}=g^{00}F_{0\nu}=\left(0,-E^{1},-E^{2},-E^{3}\right)$$

and

$$F^{\nu 0}=\begin{pmatrix}0\\E^{1}\\E^{2}\\E^{3}\end{pmatrix},\quad F^{\nu 1}=\begin{pmatrix}-E^{1}\\0\\H^{3}\\-H^{2}\end{pmatrix},\quad F^{\nu 2}=\begin{pmatrix}-E^{2}\\-H^{3}\\0\\H^{1}\end{pmatrix},\quad F^{\nu 3}=\begin{pmatrix}-E^{3}\\H^{2}\\-H^{1}\\0\end{pmatrix}.$$

For a variation of the time metric δg_{00}, we obtain

$$F^{0}{}_{\nu}F^{\nu 0}=-\vec{E}^{2},$$

which means a variation coefficient

$$F_{\mu\nu}F^{\nu\mu}\frac{1}{2}g^{00}-2F^{0}{}_{\nu}F^{\nu 0}=-\left(\vec{E}^{2}-\vec{H}^{2}\right)g^{00}+2\vec{E}^{2}=2\frac{\vec{E}^{2}+\vec{H}^{2}}{2}, \qquad \textbf{(5.62)}$$

as the double of the electromagnetic energy density

$$w^{E}=\frac{\vec{E}^{2}+\vec{H}^{2}}{2}. \qquad \textbf{(5.63)}$$

At the same time, for the variations δg_{0m} of the time-space metric elements, we obtain

$$\begin{aligned}
F_{\mu\nu}F^{\nu\mu}\frac{1}{2}g^{01}-2F^{0}{}_{\nu}F^{\nu 1} &= -\left(\vec{E}^2-\vec{H}^2\right)\underset{0}{g^{01}}-2F^{0}{}_{\nu}F^{\nu 1} \\
&= -2\left(-E^2H^3+E^3H^2\right)=2\left(E^2H^3-E^3H^2\right) \\
F_{\mu\nu}F^{\nu\mu}\frac{1}{2}g^{02}-2F^{0}{}_{\nu}F^{\nu 2} &= -\left(\vec{E}^2-\vec{H}^2\right)\underset{0}{g^{02}}-2F^{0}{}_{\nu}F^{\nu 2} \\
&= -2\left(E^1H^3-E^3H^1\right)=2\left(E^3H^1-E^1H^3\right) \\
F_{\mu\nu}F^{\nu\mu}\frac{1}{2}g^{03}-2F^{0}{}_{\nu}F^{\nu 3} &= \left[-\left(\vec{E}^2-\vec{H}^2\right)\underset{0}{g^{03}}-2F^{0}{}_{\nu}F^{\nu 3}\right] \\
&= -2\left(-E^1H^2+E^2H^1\right)=2\left(E^1H^2-E^2H^1\right),
\end{aligned} \quad \textbf{(5.64)}$$

which means that these coefficients are the double of the components of the Poynting vector

$$\vec{S}=\vec{E}\times\vec{H}\,. \quad \textbf{(5.65)}$$

For the variation of the electromagnetic potential A_μ, the variation (5.56) of the Lagrangian density of the action integral (5.54), with (5.53) and the expression of the form (3.67) of a covariant divergence, is:

$$\begin{aligned}
\delta\left(F_{\mu\nu}F^{\nu\mu}\sqrt{-g}\right) &= F^{\nu\mu}\sqrt{-g}\,\delta F_{\mu\nu}+F_{\mu\nu}\sqrt{-g}\,\delta F^{\nu\mu}=2F^{\nu\mu}\sqrt{-g}\,\delta F_{\mu\nu} \\
&= 2F^{\nu\mu}\sqrt{-g}\,\delta\left(A_{\mu,\nu}-A_{\nu,\mu}\right)=2\sqrt{-g}\left(F^{\nu\mu}\delta A_{\mu,\nu}+F^{\mu\nu}\delta A_{\nu,\mu}\right) \\
&= 4\sqrt{-g}F^{\nu\mu}\delta A_{\mu,\nu}=4\left(F^{\nu\mu}\sqrt{-g}\,\delta A_\mu\right)_{,\nu}-4\left(F^{\nu\mu}\sqrt{-g}\right)_{,\nu}\delta A_\mu \\
&= 4\left(F^{\nu\mu}\sqrt{-g}\,\delta A_\mu\right)_{,\nu}-4F^{\nu\mu}{}_{;\nu}\sqrt{-g}\,\delta A_\mu\,.
\end{aligned} \quad \textbf{(5.66)}$$

Since a total derivative has no contribution in an action integral, the total variation of the action integral (5.54), with the variations (5.56) for the metric elements and (5.66) for the electromagnetic potentials, takes the form:

$$\delta I = \int \left[\left(F_{\mu\nu} F^{\nu\mu} \frac{1}{2} g^{\rho\sigma} - 2F^{\rho}{}_{\nu} F^{\nu\sigma} \right) \delta g_{\rho\sigma} - 4F^{\nu\mu}{}_{;\nu} \delta A_{\mu} \right] \sqrt{-g} \mathrm{d}^4 x . \quad \textbf{(5.67)}$$

5.4. ELECTRIC CHARGE ACTION AND THE TOTAL MATTER-FIELD ACTION

We consider the electric current density (4.29) in the proper time $t = \tau$,

$$j^i = \dot{x}^i \rho = v^i c \rho , \quad \textbf{(5.68)}$$

and the time component of the electric current density four-vector,

$$j^0 = v^0 c \rho , \quad \textbf{(5.69)}$$

which satisfy the conservation equation (4.35),

$$\frac{\partial}{\partial x^i} j^i = -\frac{\partial}{\partial \tau} \rho = -\frac{\partial}{\partial x^0} \frac{\partial x^0}{\partial \tau} \rho = -\frac{\partial}{\partial x^0} v^0 c \rho . \quad \textbf{(5.70)}$$

With the notation (4.38), we obtain the current density four-vector of the form

$$j^{\mu} = \left(j^0 , j^i \right) = \left(v^0 , v^i \right) c \rho = 4\pi \tilde{\rho} v^{\mu} = 4\pi \tilde{j}^{\mu} . \quad \textbf{(5.71)}$$

For the normalized current density four-vector

$$\tilde{j}^{\mu} = \tilde{\rho} v^{\mu} , \quad \textbf{(5.72)}$$

the conservation equation (5.70) with (5.69) takes the form of a null covariant divergence

$$\frac{\partial}{\partial x^{\mu}} j^{\mu} = 0 . \quad \textbf{(5.73)}$$

With the charge current density four-vector (5.72), we define the charged matter action

$$I_q = -\int A_\mu \tilde{j}^\mu \sqrt{-g}\mathrm{d}^4x = -\int A_\mu J^\mu \mathrm{d}^4x. \tag{5.74}$$

With the variation of

$$J^\mu = \tilde{j}^\mu \sqrt{-g} = \tilde{\rho} v^\mu \sqrt{-g}, \tag{5.75}$$

of the form (5.36),

$$\delta J^\mu = \left(J^\nu \delta x^\mu - J^\mu \delta x^\nu\right)_{,\nu},$$

for the variation of the action (5.74),

$$\begin{aligned}
\delta I_q &= -\int \left(J^\mu \delta A_\mu + A_\mu \delta J^\mu\right)\mathrm{d}^4x \\
&= -\int \left[\tilde{\rho} v^\mu \sqrt{-g}\delta A_\mu + A_\mu \left(J^\nu \delta x^\mu - J^\mu \delta x^\nu\right)_{,\nu}\right]\mathrm{d}^4x \\
&= \int \left[-\tilde{\rho} v^\mu \sqrt{-g}\delta A_\mu + A_{\mu,\nu}\left(J^\nu \delta x^\mu - J^\mu \delta x^\nu\right)\right]\mathrm{d}^4x \\
&= \int \left[-\tilde{\rho} v^\mu \sqrt{-g}\delta A_\mu + \left(A_{\mu,\nu} J^\nu \delta x^\mu - A_{\mu,\nu} J^\mu \delta x^\nu\right)\right]\mathrm{d}^4x \\
&= \int \left[-\tilde{\rho} v^\mu \sqrt{-g}\delta A_\mu + \left(A_{\mu,\nu} J^\nu \delta x^\mu - A_{\nu,\mu} J^\nu \delta x^\mu\right)\right]\mathrm{d}^4x \\
&= \int \left[-\tilde{\rho} v^\mu \sqrt{-g}\delta A_\mu + F_{\mu\nu} J^\nu \delta x^\mu\right]\mathrm{d}^4x \\
&= \int \left[-\tilde{\rho} v^\mu \sqrt{-g}\delta A_\mu + F_{\mu\nu} \tilde{\rho} v^\nu \sqrt{-g}\delta x^\mu\right]\mathrm{d}^4x,
\end{aligned}$$

we obtain

$$\delta I_q = \int \tilde{\rho}\left(-v^\mu \delta A_\mu + F_{\mu\nu} v^\nu \delta x^\mu\right)\sqrt{-g}\mathrm{d}^4x. \tag{5.76}$$

With the electromagnetic field action (5.67) with a coefficient k, which will be determined,

$$\delta I_{em} = k\delta I = k\int \left[\left(F_{\mu\nu} F^{\nu\mu} \frac{1}{2} g^{\rho\sigma} - 2F^\rho{}_\nu F^{\nu\sigma}\right)\delta g_{\rho\sigma} - 4F^{\nu\mu}{}_{;\nu}\delta A_\mu\right]\sqrt{-g}\mathrm{d}^4x,$$

and (5.38),

$$\delta\left(I_g + I_m\right) = -\int\left[\frac{1}{16\pi}\left(R^{\mu\nu} - \frac{1}{2}g^{\mu\nu}R\right) + \frac{1}{2}\rho v^{\mu}v^{\nu}\right]\sqrt{-g}\,\delta g_{\mu\nu}\mathrm{d}^4x + \int \rho v_{\mu:\nu}v^{\nu}\delta x^{\mu}\sqrt{-g}\mathrm{d}^4x$$

the total variation of a matter-field system is

$$\begin{aligned}
\delta\left(I_g + I_m + I_{em} + I_q\right) = &-\int\left[\frac{1}{16\pi}\left(R^{\mu\nu} - \frac{1}{2}g^{\mu\nu}R\right) + \frac{1}{2}\rho v^{\mu}v^{\nu}\right]\sqrt{-g}\,\underline{\delta g_{\mu\nu}}\mathrm{d}^4x + \int \rho v_{\mu:\nu}v^{\nu}\,\underline{\underline{\delta x^{\mu}}}\sqrt{-g}\mathrm{d}^4x \\
&+k\int\left[\left(F_{\mu\nu}F^{\nu\mu}\frac{1}{2}g^{\rho\sigma} - 2F^{\rho}{}_{\nu}F^{\nu\sigma}\right)\underline{\delta g_{\rho\sigma}} - 4F^{\nu\mu}{}_{:\nu}\,\underline{\underline{\delta A_{\mu}}}\right]\sqrt{-g}\mathrm{d}^4x \\
&+\int\tilde{\rho}\left(-v^{\mu}\,\underline{\underline{\delta A_{\mu}}} + F_{\mu\nu}v^{\nu}\,\underline{\underline{\delta x^{\mu}}}\right)\sqrt{-g}\mathrm{d}^4x \\
= &-\int\left[\frac{1}{16\pi}\left(R^{\mu\nu} - \frac{1}{2}g^{\mu\nu}R\right) + \frac{1}{2}\rho v^{\mu}v^{\nu} + \frac{1}{2}E^{\mu\nu}\right]\sqrt{-g}\,\delta g_{\mu\nu}\mathrm{d}^4x \\
&+\int\left(\rho v_{\mu:\nu} + \tilde{\rho}F_{\mu\nu}\right)v^{\nu}\delta x^{\mu}\sqrt{-g}\mathrm{d}^4x + \int\left(-4kF^{\nu\mu}{}_{:\nu} - \tilde{\rho}v^{\mu}\right)\delta A_{\mu}\sqrt{-g}\mathrm{d}^4x,
\end{aligned}$$

where

$$E^{\mu\nu} = 2k\left(F_{\rho\sigma}F^{\sigma\rho}\frac{1}{2}g^{\mu\nu} - 2F^{\mu}{}_{\sigma}F^{\sigma\nu}\right)$$

is the strength energy tensor. With (5.58), from the last variation coefficient for $\mu = 0$, we obtain the third Maxwell equation (5.40),

$$\begin{aligned}
-4kF^{\nu 0}{}_{:\nu} - \tilde{\rho}v^0 &= 0 \\
-4k\nabla\vec{E} - \tilde{\rho} &= 0 \\
\nabla\vec{E} = -\frac{1}{4k}\tilde{\rho} &= 4\pi\tilde{\rho}
\end{aligned}$$

We obtain the coefficient

$$k = -\frac{1}{16\pi}.$$

We obtain the strength energy tensor

$$E^{\mu\nu} = -\frac{1}{8\pi}\left(F_{\rho\sigma}F^{\sigma\rho}\frac{1}{2}g^{\mu\nu} - 2F^{\mu}{}_{\sigma}F^{\sigma\nu}\right) \quad \textbf{(5.77)}$$

and matter-field action variation

$$\begin{aligned}\delta\left(I_g + I_m + I_{em} + I_q\right) = &-\int\left[\frac{1}{16\pi}\left(R^{\mu\nu} - \frac{1}{2}g^{\mu\nu}R\right) + \frac{1}{2}\rho v^{\mu}v^{\nu} + \frac{1}{2}E^{\mu\nu}\right]\sqrt{-g}\,\delta g_{\mu\nu}\mathrm{d}^4x \\ &+\int\left(\rho v_{\mu:\nu} + \tilde{\rho}F_{\mu\nu}\right)v^{\nu}\delta x^{\mu}\sqrt{-g}\mathrm{d}^4x \\ &+\int\left(\frac{1}{4\pi}F^{\nu\mu}{}_{:\nu} - \tilde{\rho}v^{\mu}\right)\delta A_{\mu}\sqrt{-g}\mathrm{d}^4x.\end{aligned} \quad \textbf{(5.78)}$$

From this variation, we obtain Einstein's equation of gravitation with an electromagnetic field

$$R^{\mu\nu} - \frac{1}{2}Rg^{\mu\nu} + 8\pi\rho v^{\mu}v^{\nu} + 8\pi E^{\mu\nu} = 0, \quad \textbf{(5.79)}$$

Lorentz's force in matter with a mass density ρ and a normalized charge density $\tilde{\rho}$,

$$\rho v_{\mu:\nu}v^{\nu} = \tilde{\rho}v^{\nu}F_{\nu\mu}, \quad \textbf{(5.80)}$$

and the Maxwell equations in the presence of matter:

$$F^{\nu\mu}{}_{:\nu} = 4\pi\tilde{\rho}v^{\mu}. \quad \textbf{(5.81)}$$

CONCLUSION

In the framework of the general theory of relativity, we calculated the actions of the gravitational field, mass, electric field, and electric charge, and derived the dynamic equations corresponding to the three considered variations: of the metric tensor, of the time-space coordinates, and the electric potentials. Einstein's law of gravitation in electromagnetic field, Lorentz's force acting in matter, as a function of the mass and charge densities, and generalized Maxwell equations in a gravitational field have been obtained.

REFERENCES

[1] A. Einstein (1991), *The meaning of relativity*, London, New York, Tokyo, Madras: Chapman and Hall.

[2] P. A. M. Dirac (1975), *General Theory of Relativity*, New York, London, Sydney, and Toronto: John Wiley & Sons.

[3] L. I. Schiff (1955), *Quantum Mechanics*, New York, St. Louis, San Francisco, London, Mexico, Panama, Toronto: McGraw-Hill.

[4] A. Das (2008), *Lectures on Quantum Field Theory*, New Jersey, London, Singapore, Beijing, Shanghai, Hong Kong, Taipei, Chennai: World Scientific.

[5] M. Thomson (2013), *Modern Particle Physics*, New York: Cambridge University Press.

[6] E. Stefanescu (2010), 'Master equation and conversion of the environmental heat into coherent electromagnetic energy', *Progress in Quantum Electronics* 34, 349.

[7] E. Stefanescu (2014), *Open Quantum Physics and Environmental Heat Conversion into Usable Energy*, Sharjah (UAE), Brussels, and Danvers (Massachusetts, USA): Bentham Science Publishers.

[8] E. Stefanescu (2017), *Open Quantum Physics and Environmental Heat Conversion into Usable Energy Vol. 2*, Sharjah (UAE): Bentham Science Publishers.

[9] V. Fock (1964), *The theory of space, time and gravitation*, Oxford, London, New York, Paris: Pergamon Press.

SUBJECT INDEX

Q

R

S

T

V

W